P9-AEZ-498

Industrial Application of Enzymes on Carbohydrate-Based Material

ACS SYMPOSIUM SERIES **972**

Industrial Application of Enzymes on Carbohydrate-Based Material

Gillian Eggleston, Editor
Agricultural Research Service, U.S. Department of Agriculture

John R. Vercellotti, Editor
V-Labs, Inc.

Sponsored by the
Division of Carbohydrate Chemistry

American Chemical Society, Washington, DC

Library of Congress Cataloging-in-Publication Data

Industrial application of enzymes on carbohydrate-based materials / Gillian Eggleston, editor, John R. Vercellotti, editor ; sponsored by the Division of Carbohydrate Chemistry.

p. cm.—(ACS symposium series ; 972)

Includes bibliographical references and index.

ISBN 978–0–8412–7406–8 (alk. paper)

1. Enzymes—Industrial applications—Congresses. 2. Enzymes—Biotechnology—Congresses. III. Carbohydrates—Biotechnology—Congresses.

I. Eggleston, Gillian, 1963-II. Vercellotti, John R. III. American Chemical Society. Division of Carbohydrate Chemistry.

TP248.E515 2007
660.6′34—dc22

2007060811

The paper used in this publication meets the minimum requirements of American National Standard for Information Sciences—Permanence of Paper for Printed Library Materials, ANSI Z39.48–1984.

Distributed by Oxford University Press

PRINTED IN THE UNITED STATES OF AMERICA

Dedication

This book is dedicated to my mother, Mrs. Grace Eggleston, to thank her for her encouragement and inspiration.

Foreword

The ACS Symposium Series was first published in 1974 to provide a mechanism for publishing symposia quickly in book form. The purpose of the series is to publish timely, comprehensive books developed from ACS sponsored symposia based on current scientific research. Occasionally, books are developed from symposia sponsored by other organizations when the topic is of keen interest to the chemistry audience.

Before agreeing to publish a book, the proposed table of contents is reviewed for appropriate and comprehensive coverage and for interest to the audience. Some papers may be excluded to better focus the book; others may be added to provide comprehensiveness. When appropriate, overview or introductory chapters are added. Drafts of chapters are peer-reviewed prior to final acceptance or rejection, and manuscripts are prepared in camera-ready format.

As a rule, only original research papers and original review papers are included in the volumes. Verbatim reproductions of previously published papers are not accepted.

ACS Books Department

Contents

Advances in the Applications of Enzymes in the Textile Industry

Basic Research to Underpin Future Advances in the Application of Industrial Enzymes

Indexes

Preface

The field of industrial enzyme application has grown tremendously in the past ten years, and sales are expected to be more than 2 billion U.S. dollars by 2009. This rapid growth is because of the high specificity of enzymes to catalyze industrial processing reactions, the availability of thousands of enzymes, and, especially, the rapid growth of protein engineering and genetically modified organisms to dramatically reduce production times and costs for new enzymes. Industrial enzymes also offer more environmentally friendly and scalable industrial processes for the conversion of carbohydrate-based materials into a diversity of value-added products. The new "bio-refineries" utilizing such enzymes are expected to reduce reliance on petroleum, and eventually to replace petroleum refineries. All these dramatic developments are the primary reasons to produce this book, and the associated American Chemical Society (ACS) Division of Carbohydrate Chemistry Symposium, *Industrial Application of Enzymes on Carbohydrate Based Materials,* which was organized by Dr. Gillian Eggleston. This symposium was held in August 2005 in Washington D.C. at the 230th National ACS Meeting.

The objective of this book is to provide an overall increasing awareness, understanding, and implementation of the recent great advances in the production and application of industrial enzymes on carbohydrate materials. The book focuses on carbohydrate materials as they are renewable, and the most abundant and relatively low-cost organic materials in nature such as lignocellulosic waste biomass, especially compared to traditional petroleum and gas feedstocks of ever-increasing costs. Potentially 1.3 billion tons of biomass are available as renewable feedstocks for chemicals and energy. Multiple industries are purposely highlighted so that various viewpoints can be exchanged.

The chapters are arranged so as to provide the reader with an understanding of the recent advances in industrial enzymes in four particularly fast-growing areas. The first section illustrates recent developments in the production of biofuel and industrial chemicals from underutilized carbohydrate renewable materials. The second section incorporates recent applications in the food industry that include overcoming practical problems in the technology of making and applying industrial enzymes. For the third section new innovations in the large textile industry sector are included. In the final section basic research to underpin future advances is highlighted. With contributing authors from industry, government, and academia, worldwide, the editors feel that the text should provide the readers with an up-to-date review of this exciting and rapidly developing field.

The distinguished reviewers who made this book possible by their thorough and professional reviews are acknowledged: Michael K. Dowd, John R. Robyt, Sharon V. Vercellotti, Brian Condon, Casey Grimm, Thomas Klasson, K. C. McFarland, Gregory Côté, Robert Friedman, Jin-Hee Park, Kirk Robert Hash, Sr., and Peter Wan.

We deeply appreciate the Symposium sponsors without whose donations the Symposium and book could not have been realized: ACS Division of Carbohydrate Chemistry and ACS Innovative Projects Fund Grant. Finally, we thank all the authors for their valuable contributions to this book, and we hope that the readers will enjoy it.

Gillian Eggleston
Research Chemist/Lead Scientist
Southern Regional Research Center
Agricultural Research Service
U.S. Department of Agriculture
1100 Robert E. Lee Boulevard
New Orleans, LA 70124
504–286–4446 (telephone)
504–286–4367 (fax)
gillian@srrc.ars.usda.gov (email)

John R. Vercellotti
Vice-President
V-LABS, INC.
423 North Theard Street
Covington, LA 70433
985–893–0533 (telephone)
985–893–0517 (fax)
v-labs@v-labs.com (email)

Chapter 1

Advances in the Industrial Application of Enzymes on Carbohydrate-Based Materials

Gillian Eggleston

Commodity Utilization Research Unit, Southern Regional Research Center, Agricultural Research Service, U.S. Department of Agriculture, New Orleans, LA 70124

This chapter reviews the new era of advances in the application of industrial enzymes on renewable carbohydrate materials, including mono-, di-, and polysaccharides. Stunning progress in the genetic and protein engineering of enzymes using advanced techniques of biotechnology continue to open new markets, with enzymes being increasingly tailored for specific applications. Current large-scale enzyme applications on carbohydrate materials are reported, as well as advances in their increased and cheaper production. Examples of rapid growth of the use of industrial enzymes in specific industries are described, and include fuel ethanol production from agricultural feedstocks, detergents, pulp and paper manufacture, animal feeds, fine chemicals, and specialized sugar manufacture. Practical application of enzymes under harsh industrial process conditions is still often problematic, and ways to improve the optimization of industrial applications are described. Finally, the future outlook for the application of enzymes on carbohydrate materials in industry is discussed.

Introduction

Exploitation of enzyme biocatalysts in manufacturing has occurred for thousands of years in, for example, preparation of barley for beer brewing and the leavening of bread, although these ancient processes used enzymes in whole cells. Isolated enzymes for industrial use began *circa* 1914 for use in detergents and their large-scale microbial production began in the 1960's (*1-2*). Enzyme biocatalysts remain one of the most useful industrial biotechnologies (*1*), and a new era of advances in industrial enzymes exists because of the parallel rapid growth of biotechnology. Furthermore, industrial enzymes are expected to become an even more integral part of the broad and expanding field of white technology (industrial and environmental biotechnology) that includes the use of industrial enzymes for the manufacture of bioplastics and biofuels, and abatement of environmental pollution (*3*). In a recent 2005 United Nations report, Sasson (*3*) wrote it is expected that white biotechnology and associated use of industrial enzymes "will become as ubiquitous as those of the chemical industry today because of the benefits they pose to the environment and the premium attached to this environment-friendliness for companies that undertake them."

More specific reasons for advances in industrial enzymes, particularly on carbohydrate based materials, are as follows: (i) The availability of thousands of enzymes; the number of known enzymes is estimated to be around 3000 but the number of commercialized industrial enzymes is only a little more than 100 (*4*) (ii) The high specificity of enzymes to catalyze industrial process reactions. (iii) The rapid growth of the use of protein engineering and genetically modified organisms has reduced times to produce new enzymes to weeks, dramatically increased yields, and lowered the costs of production. (iv) The environmental friendliness of industrial enzymes compared to traditional chemicals. For example, there are fewer and lower emissions including greenhouse gases and less industrial chemical waste. There is also a reduction in energy consumption by eliminating the need to maintain the typical harsh chemical environments. (v) The use of carbohydrates as feedstocks is increasing because they are the most abundant organic materials in nature, they are renewable, and because of the dramatic increases in traditional petroleum and natural gas feedstock costs in the chemical industry (*5*). (vi) The cost savings on raw materials and lower capital investment (*6*). (vii) The more efficient conversion of raw materials and standardization of the processes (*7*). (viii) The use of enzymes in more scaleable industrial processes for the conversion of carbohydrate based materials into a diversity of value-added products. (ix) The new applications, such as the bio-refineries that utilize industrial enzymes which are expected to replace petroleum refineries.

Nomenclature of Industrial Enzymes

Enzymes can be broadly classified into six major categories based on the reaction they catalyze (Table I). For a full list of enzyme classifications, the reader is referred to Dixon *et al.* (*8*). Although these six broad categories of enzymes exist in nature, hydrolases make up more than 80% of industrial enzymes (*2*), and of these approximately 50% act on carbohydrates (*9*). Hydrolases also provide most food industry related enzymes.

Table I. Broad Classification of Reactions that Enzymes Catalyze

Enzyme Class	*Reactions catalyzed*	*Examples*
Oxidoreductases	Oxidation or reduction of substrates	Dehydrogenases Oxidases Peroxidases Oxygenases
Transferases	Transfer of a group from one molecule to another	Methyltransferases Glycosyltransferases Acytransferases
Hydrolases	Bond cleavage while water is added	Esterases Glycosidases Peptidases
Lyases	Non-hyrolytic cleavage of bonds	Decarboxylases
Isomerases	Conversion of one isomer to another (intramolecular rearrangements)	Racemases Epimerases Intramolecular lyases
Ligases	Joining of two molecules at the expense of chemical energy	DNA ligase

Enzyme nomenclature was standardized in 1971 by an international agreement at a meeting of the Enzyme Commssion. Each distinct enzyme is usually assigned an EC number, and the reader is referred to Dixon *et al.* (*8*) for a full description of EC numbers. However, as well as having EC numbers, many enzymes used in industry also have both systematic and trivial names, that often cause confusion. The trivial name is often a historical derivation and does not always convey the action of the enzyme, and the systematic name can be cumbersome. In practice, it is common to refer to enzymes by their established trivial names but also include the EC number for unequivocal identification.

Current Large Scale Enzyme Applications on Carbohydrate Based Materials

Table II lists the current major large-scale applications of enzymes on carbohydrate based materials. Applications are broad and spread over a diverse range of industries (Table II). In 2002, industrial carbohydrase worldwide sales alone were US $596 million, and approximately 24% of these sales were in laundry detergents, 15% in starch-derived syrups, 14% in animal feed, 10% in textiles, 12% in pulp and paper, and 26% in all others (*10*).

Advances in the Production of Industrial Enzymes

Although some industrial enzymes are still extracted from animal or plant tissue, these are mostly proteinases. Most bulk industrial enzymes applied to carbohydrate based materials are now produced from microorganisms present in submerged cultures in fermentors (large bioreactors often $\geq$ 150,000 L capacity) for efficiency and economy. Solid state fermentation is also used for the production of some industrial enzymes, such as glucoamylase (*11*), and has been reported to offer several advantages over submerged fermentation (*11*). In this process, the solid substrate not only supplies the nutrients to the microbial culture growing in it, but also serves as an anchorage for the cells.

Figure 1 illustrates the general phases in industrial enzyme production from fermentation processes (*2*).

The first phase is the selection of an industrial enzyme. Numerous criteria are used in selection because many enzymes are applied to extreme industrial processing conditions. Criteria include specificity, reaction kinetics, optima and stability of pH and temperature, effect of inhibitors, and affinity to the substrate (*2*). Preferably, the enzyme selected must be highly active in the typical industrial concentrations of the substrate so the reaction proceeds to completion in a realistic time frame. The enzyme should also be tolerant against various heavy metals and have no need for cofactors.

In many cases the natural enzymes recruited for industrial applications do not have optimum properties for application in the industrial process as they evolved in relatively mild conditions *in vivo* compared to harsh industrial conditions. In the last two decades, there has been a two pronged approach to

Table II. Current Major Large-scale Applications of Industrial Enzymes on Carbohydrate Based Materials

Large Scale Industry	*Industrial Enzyme*	*Carbohydrate Material*	*Industrial Effect*
Detergent/Textile	Cellulase	Cellulose	Color brightening/ microfibril removal
Animal Feed	Xylanase	Xylan	Fiber solubility
Pulp and Paper	Xylanase	Xylan	To aid pulp bleaching
Starch	Amylases	Starch	Glucose formation
	Glucose isomerase	Glucose	Fructose formation (for high fructose corn syrups)
Fruit Juice	Pectinase	Pectin	Juice clarification
	Cellulase	Cellulose	Juice extraction
	Xylanase	Xylan	Juice extraction
Baking	Xylanase	Xylan	Dough conditioning
	α-Amylase	Starch	Improving loaf volume and shelf-life
	Glucose oxidase	Glucose	Dough quality
Dairy	Lactase	Lactose	Lactose hydrolysis
Brewing	Glucanase	Glucan	As a filter aid

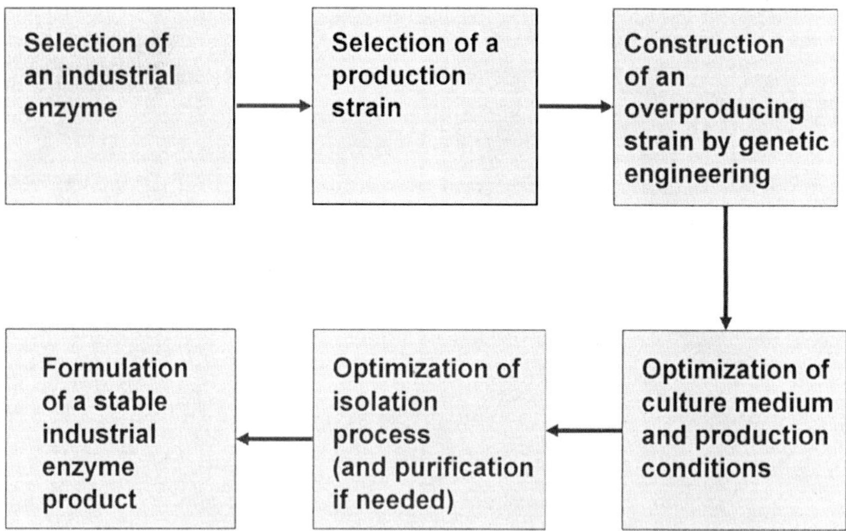

Figure 1. General scheme of process phases in industrial enzyme production.

overcome this problem: (i) discovering better enzymes in nature (bioselection), and (ii) improving a commercially available enzyme with genetic and protein engineering. Although some of the best commercial enzymes have been bioselected and expressed in benign host organisms, bioselection is often time and resource consuming. Also, sometimes when a desired property such as thermal stability has been found, other problems have occurred. Engineering of enzymes has developed rapidly over the last few years, although like bioselection, it can sometimes be time consuming. Engineering has been used to improve the yield and kinetics of an enzyme, and to clone enzymes from dangerous or unapproved microorganisms into safe, highly productive microorganisms. Many key industrial enzymes are now produced in genetically modified organisms (GMO's) (*12*). Engineering can produce and modify better enzymes in several ways, as listed in Table III.

There are now at least 30 commercial enzymes worldwide that have been engineered to function better in industrial processing (*13*). Those acting on carbohydrate-based materials include α-amylases and pullulanases (glucoamylases) used in the production of high fructose corn syrup and starch liquefaction, xylanases used in confectionary, cereals and starch, animal feed, textiles, and paper industries, and cellulases used in textiles, detergent, and animal feed industries. Some advantages for the production of enzymes from genetically modified organisms (GMO's) (*13*) are (i) Increased availability of enzymes because certain enzymes can only be produced economically by GMO methods. (ii) Enhanced efficiency. (iii) Increased purity. (iv) Enhanced process safety, *e.g.*, the use of an alternative microorganism when the original is not safe to be used in the food sector. (v) Increased economic production by reducing the process time and needed resources for the process. (vi) Increased environmental friendliness by reducing resources and waste production. One disadvantage, however, is some lack of consumer trust in enzymes from genetically modified organisms, particularly in foods.

After the industrial enzyme is selected, a production strain of the microorganism host is then chosen. Extracellular enzymes, *i.e.*, secreted from the organism, are much easier to isolate and purify.

The production host should have GRAS (Generally Regarded As Safe) status to provide regulatory assurance, which is particularly important in the food industry. The host should also be able to produce high yields of the enzyme in a reasonable time frame. Bacteria and fungi are the predominant fermentation

Table III. Genetic and Protein Engineering Techniques and Technologies to Improve the Properties of Example Industrial Enzymes.

Property Required	*Engineering Modification Procedure*	*How the Procedure Works*	*Examples*
		Genetic Engineering	
Increased productivity/ yield	**Homologous gene transfer**	The number of gene copies coding for the enzyme are increased	**Xylanase** that is used to enhance the raising of bread dough when baked (*13*)
		ProteinEngineering	
Decreased need for calcium addition	**Site-Directed Mutagenesis**	Additions/deletions/ substitutions of specific amino-acids and/or protein folds are introduced to the enzyme encoding DNA sequence	**α-Amylase** in starch liquefaction (*14*)
Increased thermal and pH stability	**Random Mutagenesis (Directed evolution)**	Mutations at random along the entire length of a gene encoding the enzyme are introduced using error-prone PCR, DNA shuffling, chemical mutagenesis, UV irradiation techniques, etc. (*14*)	**α-Amylase** in starch liquefaction (*14*)

microbial sources for enzymes acting on carbohydrates, because of their ease of use and mass production, and ability to be genetically engineered to overproduce the desired activity with no undesired side effects (*1*). In comparison, yeasts are not good producers of extracellular enzymes and, therefore, seldom used.

Once the production host has been selected and/or genetically engineered to overproduce the desired enzyme, a fermentation production process is developed (Figure 1). Optimization of the growth conditions includes media composition, cultivation type, and process conditions; it is a demanding task that often involves as much effort as genetic engineering of the host (*2*) (Figure 1). Isolation or recovery of the enzyme product occurs after cell removal from the fermentor by vacuum drum filtration, separators, or microfiltration (*2*). Purification of the isolated enzyme may then be required by ion exchange or gel filtration techniques. The isolated enzyme is then preserved and formulated (Figure 1). Industrial enzymes are often formulated as liquids with preservatives added such as salts or polyols, or can be formulated as granules or a non-dusty dry product.

Practical Application of Enzymes in Industrial Processing

Application of enzymes to industrial processes is often problematic and difficult to optimize because conditions are usually harsh. Also, extrapolation from performance in the laboratory and pilot plant scale to the industrial scale is not always linear. The simplest way to apply enzymes is to add them directly to the process stream, as a liquid or solid enzyme formulation, where they catalyze the desired reaction and are gradually inactivated downstream of the process. Deactivation is especially important in the food industry, where enzymes activities should not be retained in the end products. Direct addition of an enzyme is often the case in numerous bulk enzyme applications, *e.g.*, bleaching of wood pulp with xylanases. In bulk applications the price of the enzyme must be low to make their use economical (*2*), and cost of the enzyme relative to that of other methods determines the demand for the enzyme.

An alternative way to apply enzymes is to immobilize them on a solid support so that they can be re-used. There are different methods to immobilize enzymes (*15*): covalent attachment to a solid support, entrapment in a gel matrix such as agar, adsorption onto an insoluble matrix, and intermolecular cross-linking of the enzyme to form an insoluble matrix. Advantages of immobilization are continuous output of products, control of the reaction rate by regulating flow rate, easy stoppage of the reaction for the facile removal of the enzyme, non-contamination of the product, and improved enzyme stablility. Production of HFCS (high fructose corn syrup) has been greatly facilitated by the use of immobilized glucose isomerase, allowing continuous HFCS production. However, immobilized enzyme systems can present significant industrial problems, particularly with mass transfer. They generally require the substrate to be a relatively small molecule, *i.e.*, glucose in HFCS production, because the substrate and resulting product must be able to easily move in and out of the support matrix. Immobilized enzyme systems are also expensive, and to be cost effective in many industries have to be used for at least one year or more, to pay off the initial capital cost. A more promising technology to improve the efficiency of industrial enzymes is the use of ultrasound technology, but this has not yet been commercialized (*16-17*).

In many industries, *e.g.*, the sugar (*18-19*) and animal feed industries (*20*), there have been no uniform/standard methods to measure the activity of commercial enzymes sold to users by vendors or distributors. This has meant that direct comparison of activities by users has not been possible and, unfortunately, there is no regulatory body in the U.S. to issue or regulate standard activity method and unit for a commercial enzyme. In the U.S. sugar industry, very wide variations (up to 20-fold) in activity of commercial dextranases have been reported (*21*) that do not always reflect the costs of the enzymes. This has greatly reduced the efficiency of the enzyme in some factories because the users were unknowingly adding an enzyme of low activity. Eggleston and Monge (*18*) reported the urgent need for a standardized method to measure the activity of commercial dextranases at sugarcane and sugarbeet factories and refineries to (1) compare economically equivalent activities of different dextranases, (2) measure the activity of delivered batches, and (3) monitor the changing activities on factory storage, and they introduced a simple factory method. Marquardt and Bedford (*20*) have also stated that, for any industry, there should be a correlation between the activity values obtained from the standardized method and its general effects in the industrial process. Furthermore, because in some industries enzymes are usually sold to users by small vendors or distributor companies (who purchase the enzymes in bulk quantities from much larger enzyme producer companies), they often cannot afford the same application support as the larger companies. Their sales and technical staff often do not have the expertise or knowledge found in the large enzyme companies to aid users to optimize their enzyme applications.

Another reason for difficulties in optimizing certain industrial enzyme is the amount of research and development put forth to optimize the enzyme's efficiency in the harsh industrial environment. For enzymes with large markets and high volume sales, *i.e.*, α-amylases used in HFCS production, baking and numerous food industry processes, a great deal of R and D has been expended to optimize and tailor the properties of the enzyme for the specific application, *e.g.*, the engineering of α-amylases to be highly temperature stable and have minimum calcium requirements in HFCS production. For enzymes with relatively small markets and low volume sales, very limited R and D has been undertaken to optimize the enzymes properties for the industrial conditions.

These include "band-aid" enzymes that are only used when there is a processing problem; for example, in the sugar industry, application of dextranase occurs only when sugarcane or sugarbeet deterioration has occurred.

The Use of Industrial Enzymes in Ethanol Production from Agricultural Feedstocks

The chemical industry has been based on a petroleum-based menu for decades. However, current economic, political, and environmental factors are now driving a trend toward greater use of agricultural bio-based feedstocks, which increasingly include renewable carbohydrate-based materials to produce ethanol as an alternative biofuel.

Ethanol, once viewed as a niche fuel extender, is now accepted worldwide as an important liquid fuel, valued for both its oxygen content and ability to raise the octane rating (*22*). It can be produced from agricultural products and/or waste feedstocks containing starch, sucrose, and even cellulose from, for example, corn, corn stover, or sugarcane bagasse. Fuel ethanol production and its utilization is on the rise worldwide, particularly in North and South America (*22*), with Brazil the world leader. The European Union is currently developing a coherent biofuel strategy. The challenges presented on the world stage are that most ethanol-fuel programs enjoy subsidies, tax breaks or both, and that large-scale imports could potentially disrupt the cycle of local production. Most ethanol is currently being consumed by the country producing it, although this is expected to change (*22*). Nevertheless, the world's first biorefinery by Iogen Corporation is now fully operational as a demonstration plant in Canada (*6*), and is converting cellulosic material, *i.e.*, wheat and other cereal straws and corn stover, into fuel ethanol for blending with petroleum gasoline. To reduce overall conversion costs, the cellulose feedstock is pre-treated with a modified steam explosion process to increase the surface area of the plant fiber available to enzymes. Highly potent and efficient cellulase enzyme systems, tailored to the specifically pre-treated feedstock, are used in multi-stage hydrolysis reactors to break down the cellulose to glucose for fermentation into fuel ethanol. The separated lignin in the plant fiber is also used to drive the process by generating electicity and reduce fossil fuels usage (*23*).

In the future, production costs for fuel ethanol are expected to fall. Cellulose enzyme costs are no longer the main economic barrier to the commercialization of cellulose-fuel ethanol processes. In 2005, Novozymes Corporation reported (*24*) that, with research grants from the U.S. Department of Energy and the National Renewable Energy Laboratory, cellulose-ethanol production costs were reduced 30-fold since 2001 from "US$5.50/gallon ethanol to US$0.10-0.18, exceeding their initial goal of a 10-fold cost reduction. This was accomplished through an integration of bioinformatics, proteomic and microarray analyses, and a directed enzyme evolution program for the creation of a single cellulase-producing organism with significant improvements in enzyme yield, activity, and thermostability, as well as improvements in feedstock pretreatment. The reduction in enzyme cost is based on laboratory-scale conversion of acid-pretreated corn stover, so translation to commercial scale or to other feedstocks will alter the enzyme cost (*25*).

Enzymes in Detergents

The use of industrial enzymes in detergent formulations is common in developed countries, with over half of all available detergents containing enzymes. In spite of the detergent industry representing the largest single market for industrial enzymes at 25-30% of total sales, details of the specific enzymes used and how they are used have seldom been published. The use of enzymes allows low temperature, short duration, and mechanical energy approaches to washing. Enzymes are used in small amounts in most detergent formulations, *i.e.*, only 0.4 – 0.8% crude enzyme by weight (~1% by cost). Currently only proteases and amylases are commonly used (often together in a single formulation), and are all produced by *Bacillus* bacteria. α-Amylase, used in detergents to remove starch based stains, is from *Bacillus licheniformis* and is the same as that used in the production of glucose syrups (*26*). α-Amylase is active at temperatures of 85 °C and can tolerate pH values ~10.5, and is especially used in dish-washing and de-starching detergents. Recently, alkaline-stable fungal cellulases have been introduced into detergents used for washing cotton fabrics (*27*). The cellulases remove microfibrils and pills from the surface of the material (for color brightening and softening), and aid in the removal of soil particles from the wash by hydrolyzing associated cellulose fibers. The use of enzymes in washing detergents has still not reached its full potential, and more enzymes targeted to a specific stain or problem are expected in the future.

Increased Use of Industrial Enzymes in Pulp and Paper Manufacture

In the last two decades, enzymes have established niche roles in a number of areas in the pulp and paper industry. The reasons for this are the increased availability of a whole range of enzymes at reasonable cost, made-to-order enzymes based on genomic information from the major wood-degrading microorganism now publically available, and the large research effort by numerous industrial enzyme companies to develop a cost-effective portfolio of enzyme-based applications on papermaking carbohydrates (*23, 28*).

Since 1984 (*29*), there has been widespread addition of xylanase to pulp brownstock prior to bleaching to save on bleaching chemicals, although the mechanism of action is still not fully delineated (*30*). In Japan, xylanase is even being produced on site at a mill; from a bacterial fermentation of a pulp side-stream which results in a xylanase/pulp mixture. More recently, cellobiohydrolase, part of the family of cellulases, and pectinase have been added to modify carbohydrates in pulp wood chips or pulp reject fractions (biopulping) to save on mechanical pulping energy (*31-32*). However, these applications have yet to be used large-scale. In numerous mills, particularly those using a starch-based coating system, amylase in combination with lipase and protease is being effectively used in paper machine boil-outs. Results are more favorable than traditional caustic chemical treatment (*23*). Amylase has also been effective in removing and controlling the growth of bacteria in paper machine systems. There have been some trials on the use of cellulases to facilitate de-inking of secondary (recycled) cellulose paper and waste. However, cellulases require neutral or acid pH conditions, which have made them difficult to compare with conventional alkaline de-inking chemistries (*33*). The trend of de-inking chemistry towards neutral conditions, however, should allow more opportunity to use cellulases in de-inking plants (*23*).

Industrial Enzymes Action on Animal Feed Carbohydrates

Industrial enzymes now play a crucial role in animal feeds worldwide. Enzymes that act on carbohydrates in feeds are mainly designed to degrade components of the raw feed materials that limit digestibility, and thus improve feed digestability (non-starch polysaccharides or fiber), while making it possible to use cheaper raw materials. Furthermore, enzyme supplementation of cereal-based diets such as wheat, barley, wheat, rye, and oats, has been instrumental in producing more uniform performance values in poultry and other livestock (*20*). The first commercial success was addition of β-glucanase into cheaper barley-based feeds by Finnfeeds International Ltd. Barley contains β-glucan that causes high viscosity in chicken guts. β-Glucanase causes increased animal weight gain with the same amount of barley. Xylanases are now routinely used in many wheat-based feeds, and many feed-enzyme preparations are now multiple enzyme-flour mixtures containing glucanases, xylanases, proteinases, and amylases. Cellulases, hemicellulases, pectinases, and phytases are other enzymes also added sometimes.

Although enzyme feed additives have had a great impact on the animal livestock industry, the use of enzymes is still in progress. Many problems still need to be solved before their full potential is reached. These include the need for standard enzyme activity assays for assessing the activity and quality of commercial enzyme products (*20*). New forms of enzymes with better properties are needed, including high activity under process conditions, high levels of resistance to inactivation by heat, low pH, and proteolytic enzymes, low production costs, and long shelf-life under ambient storage conditions (*20*). The biotechnology gene techniques discussed in the section titled Advances in the Production of Industrial Enzymes in this book chapter, will most likely be used to produce these needed enzymes from genetically modified organisms.

Industrial Enzymes in the Production of Fine Chemicals and Specialized Sugars

Other main users of industrial enzymes are the Fine Chemical industries, where volumes of enzymes used are relatively small but prices are high. Enzymes have been used in the manufacture of fine chemicals for a long time, for example in beverage alcohols, but the whole living cell has usually been used (*2*) because of the need to combine energy sources from ATP or NADPH to the process. However, isolated enzymes have been successful in some fine chemical synthesis such as the production of rare sugars and oligosaccharides. Rare, non-natural monosaccharides are needed as starting materials for new chemicals and pharmaceuticals, e.g., D-psicose, L-xylose, and D-tagatose. Recently, enzymes have been developed to manufacture practically all D- and L-forms of simple sugars (*2*). Glucose isomerase, one of the most important industrial enzymes for utilization of carbohydrate materials, is used in the production of high fructose corn syrup, a frequently used feedstock in the beverage and baking industries.

Enzymes are also being used to synthesize oligosaccharides, as chemical syntheses of oligosaccaharides are extremely complicated. Typical enzymes used are glycosyltransferases that catalyze the transfer of monosaccharides from a donor to a sugar acceptor. For example, glucansucrases for the synthesis of novel oligosaccharides (*34*). Hydrolases are also being used to produce oligosaccharides from the breakdown of polysaccharides (*35*). The best known and most widespread use of hydrolases is undoubtedly the hydrolysis of starch, which is covered in other chapters of this book. Another example is the use of *endo*-mannanase to covert guar gum to a mixture of low-molecular weight galactomannan oligosaccharides, that are used as a soluble dietary fiber and prebiotic.

Other Sources and Roles

Syngenta corporation (*36*) has recently developed technology to produce a variety of novel starch hydrolyzing enzymes directly in corn grain, enabling the design of corn varieties that are tailored to specific food, feed, and processing applications. This development will allow the flexible and high level enzyme expression in a dry, stable, pre-packaged formulation. The genetically modified corn has the same starch, protein, and oil contents as its non-genetically modified counterparts, but it still awaits approval as food and feed.

Enzymes are also used in analytical methodology, such as the use of mannitol dehydrogenase in the measure of mannitol in urine (*37*) to diagnose intestinal permeability problems and in sugarcane as a measure of deterioration (*38*). Enzymes in analytical methods have to be very pure and free from interfering side activities. Some analytical enzymes are now being immobilized on electrodes and commercial instruments are available.

Enzymes used on carbohydrate materials also occur in personal care products. For example, some toothpaste contain glucoamylase and glucose oxidase to break down starch based products into disinfectant compounds (*2*).

Future Outlook

The future of application of industrial enzymes on carbohydrate materials looks bright especially for use in bio-manufacturing processes. Petroleum continues to be replaced by polysaccharides as the renewable, biomass feedstocks for the new biorefineries. The use of cellulases to convert waste cellulose from various crops into fermentable sugars for fuel ethanol production has been a major study topic in recent years, and the cost of these enzymes is no longer the main economic barrier to the commercialization of this technology (*25*). Increasing environmental pressures and energy prices will make large scale application a real possibility in the near future (as well as reducing cost of enzyme production).

Enzyme based biocatalytic reactions are also a key technology for "green chemistry" and the natural solutions to many industrial problems. As a consequence, expectations for the use of enzymes are wide-ranging over various other fields including cleaning the environment, re-using resources (recycling), creating new industrial processes, creating new functional foods, and contributing to medical treatments. Furthermore, adoption of biobased processes utilizing industrial enzymes is gaining momentum because of their potential both for significant improvements in process profitability and for massive market growth in the future (*5*).

In the future, more enzymes will be commercially available for industrial use from today's current "enzyme pool", discovery of more and more naturally occurring enzymes with unique properties adaptable to industrial use, and the use of genetic and protein engineering (*1*). There will be more specifically tailored enzymes for applications, *e.g.*, there will be a specifically tailored xylanase for baking, another one for feed, and a third for pulp bleaching (*2*). It could be expected that some enzymes may be redesigned to fit more appropriately into other industrial processes. Some enzymes are expected even to be manufactured directly in the crop plant (*36*). Continued research on the synthesis of artificial enzymes (synzymes) may also be expected to produce revolutionary enzymes that act on carbohydrate materials (*1*).

Optimization of the practical applications of enzymes in industrial processes is still required in certain industries, particularly industries relying on enzymes with low sale volumes that have had little research and development to improve their performance under the desired industrial conditions.

References

1. Lesney, M. S. *Today's Chemist at Work* **2003**, Dec. edn., p 20-23.
2. Leisola, M.; Jokela, J.; Pastinen, O.; Turenen, O.; Schoemaker, H. *Biotechnol.* www.hut.fi/Units/Biotechnology/Kem-70.415, **2001,** 1-25.
3. Sasson, A. In *Industrial and Environmental Biotechnology: Achievements, Prospects, and Perceptions*. **2005**, UNU-IAS Report.
4. Lievonen, J. In *Technological Opportunities in Biotechnology*, Technical Report of VTT (Technical Research Center of Finland), **1999**, No. 43/99.
5. Kerr, E. *Genetic Eng. News* **2004**, *24(6).*
6. Riese, J. *Summary Proc. World Congress on Industrial Biotech. and BioProc.* Orlando, FL **2004**, April 21-23, p 6.
7. Taylor, A.J.; Leach, R.M. In *Enzymes in Food Processing*, Tucker, G.A.; Woods, L.F.J.; Eds.; Blackie Academic and Professional Press: London, **1995**, p 26-40.
8. Dixon, M.; Webb, E.C.; Thorne, C.S.R.; Tipton, K.F. In *Enzymes*, 3rd edn., Longman: London, **1979**.
9. Righelato, R.C.; Rodgers, P.B. In *Chemical Aspects of Food Enzymes*, Andrews, A.T. Ed.; Royal Society of Chemistry: London, **1987**, p 289-314.
10. Anon. Business Communications Company, Inc. Report, **2004**, Figure 14.
11. Pandey, A.; Selvakumar, P., Soccol, C. R.; Poonam, N. *Current Sci. (Bangalore)*, **1999**, *77(1)*, 149-162.
12. Faigan, C. O. *Enzyme Microb. Technol.* **2003**, *33*, 137-149.
13. Anon., *www.europabio.org/module_14*, **2006**, p 1-9.
14. Hashida, M.; Bisgaard-Frantzen, H. *Trends Glycosci. Glyc.*, **2000**, *12*, 389-401.
15. Worsfield, P. J. *Pure Appl. Chem.*, **1995**, *67(4)*, 597-600.
16. Yachmenev, V; Blanchard, E.; Lambert, A. *Ultrasonics* **2004**, *42*, 87-91.
17. Yachmenev, V.; Condon, B. D.; Lambert, A. In *Industrial Application of Enzymes on Carboydrate Based Materials*; Eggleston, G.; Vercellotti, J. R.; Eds.; ACS Symp. Series: Washington D.C. **2007**, in press.

18. Eggleston, G.; Monge, A. *Process Biochem.*, **2005**, *40(5)*, 1881-1894.
19. Eggleston, G.; Monge, A.; Montes, B. In *Industrial Application of Enzymes on Carboydrate Based Materials*; Eggleston, G.; Vercellotti, J. R.; Eds.; ACS Symp. Series: Washington D.C. **2007**, in press.
20. Marquardt, R. R.; Bedford, M. In *Enzymes in Poultry and Swine Nutrition*, Marquardt, R. R.; Zhengkang, H.; Eds.; **1997**, IDRC publication, p 154.
21. Eggleston, G.; Monge, A.; Montes, B.; Stewart, D. *Intern. Sugar J.*, **2006**, in press.
22. Cezanne, J. *Here* **2004**, July edn., Alfa-Laval publication, p 4-6.
23. Anon. *www.iogen.ca*, **2005**.
24. Anon. *www.novozymes.com*, **2005**.
25. Teter, S. *Summary Proc. World Congress on Industrial Biotech. and BioProc.* Orlando, FL **2004**, April 21-23, p 13-14.
26. Chaplin, M. In *The Use of Enzymes in Detergents*, *www.lsbu.ac.uk*, **2004**, p 1-4.
27. Ito, S. *Extremophiles* **1997**, *1(2)*, 61-66.
28. Kenealy, W.R.; Jeffries, T. W. In *Wood Deterioration and Preservation: Advances in Our Changing World*, Goodell, B.; Nicholas, D.D.; Eds.; ACS Symposium Series 845: Washington D.C. **2003**, p 210-239.
29. Viikari, L.; Ranua, M.; Kantelinen, J.; Sundquist, J; Linko, M. *Proc. Third International Conf. on Biotech. Pulp Paper Industry*, Stockholm, **1986**, p 67-69.
30. Paice, M. G.; Renaud, S.; Bourbonnais, S.; Labonte, S.; Berry, R. *J. Pulp Pap. Sci.* **2004**, *30(9)*, 241-246.
31. Pere, J.; Ellmen, J.; Honkasalo, P.; Taipalus, P.; Tienvieri, T. In *Biotechnology in the Pulp and Paper Industry*, 8th ICBPPI, Viikari, L; Lantto, R.; Eds.; Elsevier: Amsterdam, **2002**, p 281-290.
32. Peng, F.; Rerritsius, R.; Angsas, U. *Internat. Mech. Pulping Conf. Proc*, **2003**, p 335-340.
33. Xia, Z.; Beaudry, A.; Bourbonnais, R. *Progress Paper Recycling* **1996**, *5(4)*, 46-58.
34. Monsan, P.; Paul, F. *FEMS Microbiol. Rev.*, **1995**, *16*, 187-192.
35. Eggleston, G.; Cote, G, L. In *Oligosaccharides in Food and Agriculture,* G. Eggleston, G.; Cote, G. L.; Eds.; ACS Symp. Ser. 849: Washington D.C. **2003**, p 1-14.
36. Lanahan, M. *Summary Proc. World Congress on Industrial Biotech. and BioProc.* Orlando FL, April 21-23, **2004**, p 13-14.
37. Lunn, P. G.; Northrop, C. A.; Northrop, A. *Clin. Chim. Acta* **1989**, *183*, 163-170.
38. Eggleston, G.; Harper, W. *Food Chem.* **2006**, *98*, 366-372.

Advances in the Application of Industrial Enzymes in Biofuel and Industrial Chemical Production

Chapter 2

Development of Improved Cellulase Mixtures in a Single Production Organism

K. C. McFarland, Hanshu Ding, Sarah Teter, Elena Vlasenko, Feng Xu, and Joel Cherry

Novozymes, Inc., 1445 Drew Avenue, Davis, CA 95616

Economical conversion of lignocellulosic biomass to sugars, ethanol or other chemical feedstocks requires an optimization of the enzymatic breakdown of plant cell walls to monosaccharides. Corn Stover was converted to glucose by dilute acid pretreatment followed by enzymatic digestion with broths containing novel mixtures of proteins expressed in *Trichoderma reesei*. Optimization included improvement of the pretreatment, alteration of the expressed suite of enzymes, and increased productivity of the expression host, which resulted in significant reduction in the enzyme cost for utilization of lignocellulosic biomass for the production of fuels and chemicals.

The development of commercially viable renewable chemicals and energy feedstocks is of economic, environmental, and strategic importance. Fuel ethanol production from starch and sugar biomasses, such as corn grain and sugarcane is rapidly expanding, with North American production for 2005 estimated at nearly 20 billion liters, with a projected annual growth rate of ~20%. However, the 13.5 million tons of corn grain currently used in the U.S. for ethanol production is dwarfed by the 1.3 billion tons of biomass potentially available as a renewable feedstock for energy and chemicals (*1*). The conversion of lignocellulosic biomass to sugars, ethanol, and other chemicals holds the promise of the ability to reduce pollution, add value to agriculture production, and reduce reliance on petroleum. We describe the removal of a significant barrier to use of these feedstocks by development of a very efficient cellulase system expressed at industrially useful levels in a single organism.

The most abundant agricultural biomass is corn stover (corn leaves and stalks): current U.S. cultivation produces about 225 million tons of corn stover, of which about 90 million can be sustainably harvested. Intensive studies on conversion of corn stover to sugars and ethanol are under way, with one method comprised of a pretreatment by heat and dilute acid prior to enzymatic hydrolysis to produce sugars that can then be fermented to ethanol or used for the synthesis of other high value products. During a multi-year joint effort with the National Laboratory for Renewable Energy (NREL) funded by the Department of Energy, we investigated ways to significantly enhance the conversion of corn stover to fermentable sugars and ethanol, focusing on improvements in the enzymatic hydrolysis of pretreated corn stover (PCS). Using the *Trichoderma reesei* cellulolytic system as a starting point, we systematically studied the interaction between cellulase and lignocellulose. Various methodologies, including directed evolution, molecular biology, DNA microarray, proteomics, biochemistry, bioinformatics, and high throughput robotic assays were applied. We evaluated the enzyme production of various cellulolytic fungi, probed the structure-function relationship of cellulases, engineered thermally active mutants, created synergistic cellulase mixes, and significantly enhanced enzymatic corn stover hydrolysis. Our goal was to decrease enzyme production costs and increase enzyme activity, with improved enzyme mixtures produced in a single high-expression organism.

Enzymatic hydrolysis of cellulose has been studied for decades, and the enzymes are classifiec based on sequence homology into glucoside hydrolase families 1, 3, 5-9, 12, 44, 45, 48, 61, and 74 (*2*). Based on the mode of action, cellulases can be grouped into exo-1,4-β-D-glucanases or cellobiohydrolases (CBHs, EC 3.2.1.91), endo-1,4-β-D-glucanases (EGs, EC 3.2.1.4) and β-glucosidases (BGs, EC 3.2.1.21). One of the most studied fungal cellulolytic systems is that of *Trichoderma reesei* (*3*); it is one of the most effective cellulose degraders, and currently the basis of commercial cellulase products such as Novozymes' Celluclast 1.5L (Celluclast). The wide array of cellulases in these products is capable of converting almost all of the glucan in PCS to sugars, however the cost for effective dosage was too high for an economically viable biomass-to-ethanol process (*4-5*).

Enzymatic conversion of cellulosic feedstocks is a crucial step in a multi-step process. After collection and delivery, and often storage of the biomass, is pre-processing to remove foreign, non-plant material, combined with physical size reduction by chopping or milling. Pretreatment follows, entailing thermal and/or chemical treatment that increases the accessibility of the cellulose in the biomass to enzymatic hydrolysis. The result of an effective pretreatment is a dramatic reduction in the amount of enzyme (cellulase) required to convert the cellulose polymer to glucose and an increase in the yield of fermentable sugars. Both of these effects have a significant impact on the overall process economy (*6-7*). There are a number of promising and proven pretreatment technologies available at varying levels of implementation, but the dilute acid method has been most thoroughly studied (*8-9*). Here, preprocessed stover is incubated for 1-4 min at 190-198 °C at 12.1 atm pressure, effectively hydrolyzing the hemicellulose fraction to monomeric, soluble sugars (*4,10*). Composition of stover is 30-40% cellulose, 30-40% hemicellulose, and 15-20% lignin. After dilute acid pretreatment, the solid fraction contains ~57% cellulose, ~5% hemicellulose, and ~28% lignin, while the pentose sugars released from the hemicellulose are solubilized in the liquid fraction. Improvements in pretreatment and resulting improvements in digestibility were shown over the term of the collaboration. Standard digestion conditions are described in NREL LAP008 and LAP009 (*11*), and briefly summarized were a 50 g mixture comprised of 6.7% cellulose (approximately 13% washed PCS dry solids), 50 mM citrate or acetate buffer, pH 5.0, and enzyme dosing from 1 to 30 FPU per g cellulose, incubated at 50 °C with orbital shaking for 1 to 7 days, with analysis of sample aliquot supernatants. Numerous miniaturized assays and analog substrates (e.g. phosphoric acid swollen cellulose or PASC, cellobiose, carboxymethyl cellulose or CMC, 4-methylumbelliferyl-beta-D-lactoside or MUL) were also used. Hydrolysis of these substrates was measured as reducing sugars by reaction with para-hydroxy benzoic acid (*12*), by spectroscopy, or integrated refractive index detection after separation of hydrolysates by high performance liquid chromatography using an Aminex HPX-87H column (BioRad) eluted with 5 mM sulfuric acid at 65 °C. The chosen benchmark cellulolytic activity was Celluclast (~60 FPU/mL) supplemented with 1% w/w Novozym 188, containing our *Aspergillus niger* BG (~404 IU/mL) product.

The cellulolytic system of *Trichoderma reesei* includes CBHI (~60%), CBHII (15%), EGI and –II (~20%) and other minor components, and it secretes these cellulases at high levels under appropriate induction. These enzymes act synergistically to hydrolyze cellulose, so the selection and improvement of each functional family is likely to improve the overall hydrolysis. The optimum pH and activity for these enzymes vary slightly, but are generally near pH 5 and 50

°C. The activity of this system is not fully stable at temperatures above 55-60 °C. In order to produce a cellulase system that would work efficiently at various conditions that may occur in a biorefinery, we applied the techniques of directed evolution to the processive cellobiohydrolase CBHI, the most abundant cellulase, for the purpose of increasing its temperature stability and, in turn, the rate of reaction at higher temperatures (*13*). Repeated cycles of gene mutation by error prone PCR, subcloning into a *Saccharomyces cerevisiae* expression system, robotic screening for selection of improved characteristics followed by recombination of mutations by gene shuffling to generate libraries suitable for further screening, allowed us to select candidate mutants. We succeeded in producing variants capable of higher activity relative to wild-type CBHI at elevated temperatures of 60-70 °C. The results, reported as the ratio of activity on the fluorescent substrate MUL of each variant at the high test temperature divided by the activity at 50 °C distinguish mutants of increased stability. A screen at 63 °C and 50 °C shows mutants with ~6-fold higher activity than wild-type CBHI (Figure 1). More than a 16-fold improvement in stability of activity of the enzyme was achieved at 70 °C (data not shown). Resulting variants were shown to have improved performance in the long term (5-7 day) hydrolysis of PCS at 60 °C, that was higher when combined with thermotolerant CBHII, EG and BG. However, these mutants have lower activity at 60 °C or above than wild type CBHI at 50 °C.

Directed evolution of the *Aspergillus oryzae* Family 3 β-glucosidase increased the relative stability of BG under heat stress at elevated temperatures (e.g. 65 °C), but at the cost of reduced activity at 50 °C compared to wild type, with activity at 65°C approximately equivalent to that of wild type at 50 °C (*14*).

Increasing number of EG have been studied in terms of structure and reactivity, but we expanded this to compare ~20 EG from families 5, 7 and 45 with various substrates (Figure 2). All families showed cellulase activity on phosphoric acid swollen cellulose (PASC). Family 5 EG also showed mannanase activity, while family 7 EG showed xyloglucanase activity. Family 45 EG did not show significant activity on xyloglucan, mannan or p-nitrophenyl-β-D cellobioside. Awareness of the substrate specificity for these family members allows us the ability to tailor cellulase systems to the composition of specific biomasses with different compositions of cellulose and hemicellulose (*15*).

Cellulolytic microorganisms express many varied proteins in response to growth on cellulosic carbon sources. With the goal of enhancing the cellulolytic capabilities of *Trichoderma*, we used molecular biology methods including Expressed Signal Tags (EST) and DNA microarray analysis of several organisms. Microarray expression analysis allows the comparison of large numbers of mRNA levels in an organism, which can be correlated to the differential expression and secretion of proteins important for growth on different carbon sources (*16*). For example, we grew *Trichoderma* cultures on glucose and on PCS, and cDNA probes were made from their respective mRNAs. Hybridization of cDNA probes to Suppression-Subtractive-Hybridization libraries spotted on microarrays allowed detection of 728 clones specifically induced by growth on PCS vs. glucose. Analysis of a genomic array led to an additional 170 PCS-induced clones (*17*). Sequence analysis showed about 90% had canonical fungal signal sequence. Each was analyzed to determine if they might enhance cellulolytic activity.

Concurrently, we probed our large microbial collections for broths containing proteins capable of boosting the activity of *T. reesei* cellulases. One such example was broth from another cellulolytic fungus *(Fungus Z)*, which

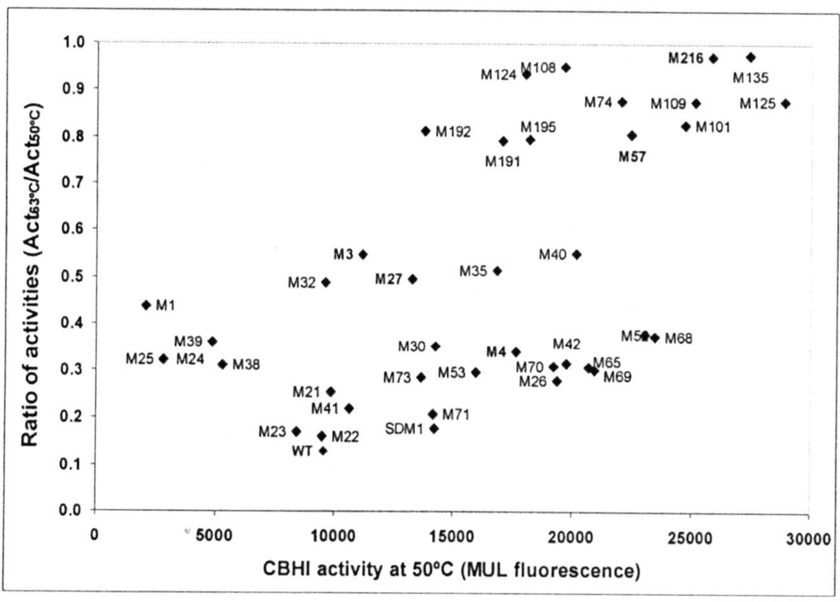

Figure 1. Yeast-expressed broths containing mutant CBHI was tested on MUL at 63 °C and 50 °C in a screen designed to highlight mutants with increased stability of activity at higher temperatures. The activity at the lower temperature (X-axis) is the raw fluorescent signal obtained. The ratio of activity (Y-axis) is the ratio of fluorescence obtained at 63°C divided by fluorescence obtained at the 50 °C. For example, Wild-type CBHI (marked WT) shows MUL fluorescence signal of ~9000 at 50 °C and about 0.15 of that activity at 63 °C. The quantity of protein in the broths was not normalized, and the activity of yeast-expressed vs. Trichoderma-*expressed proteins was not determined. However the prediction of high temperature stability was confirmed for many mutants in this screen.*

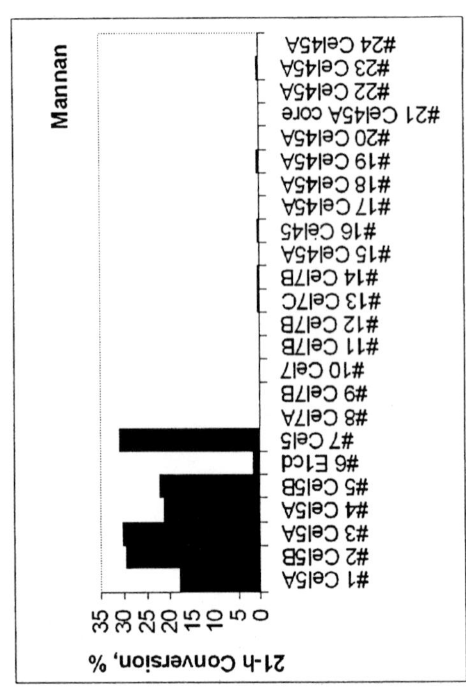
Mannan
21-h Conversion, %
0
5
10
15
20
25
30
35
#1 Cel5A
#2 Cel5B
#3 Cel5A
#4 Cel5A
#5 Cel5B
#6 E1cd
#7 Cel5
#8 Cel7A
#9 Cel7B
#10 Cel7
#11 Cel7B
#12 Cel7B
#13 Cel7C
#14 Cel7B
#15 Cel45A
#16 Cel45
#17 Cel45A
#18 Cel45A
#19 Cel45A
#20 Cel45A
#21 Cel45A core
#22 Cel45A
#23 Cel45A
#24 Cel45A

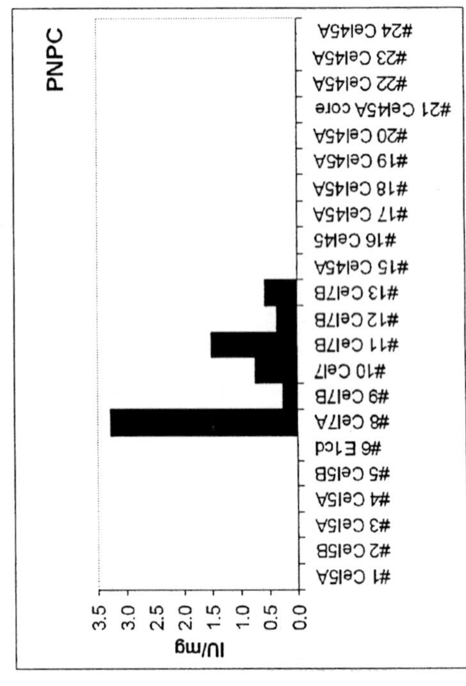
PNPC
IU/mg
0.0
0.5
1.0
1.5
2.0
2.5
3.0
3.5
#1 Cel5A
#2 Cel5B
#3 Cel5A
#4 Cel5A
#5 Cel5B
#6 E1cd
#8 Cel7A
#9 Cel7B
#10 Cel7
#11 Cel7B
#12 Cel7B
#13 Cel7B
#15 Cel45A
#16 Cel45
#17 Cel45A
#18 Cel45A
#19 Cel45A
#20 Cel45A
#21 Cel45A core
#22 Cel45A
#23 Cel45A
#24 Cel45A

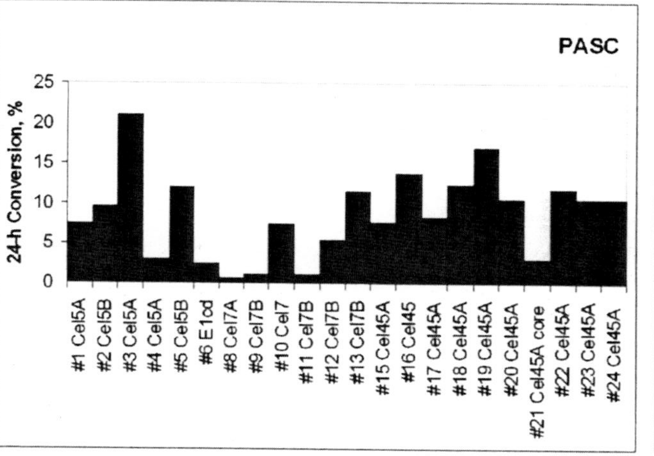

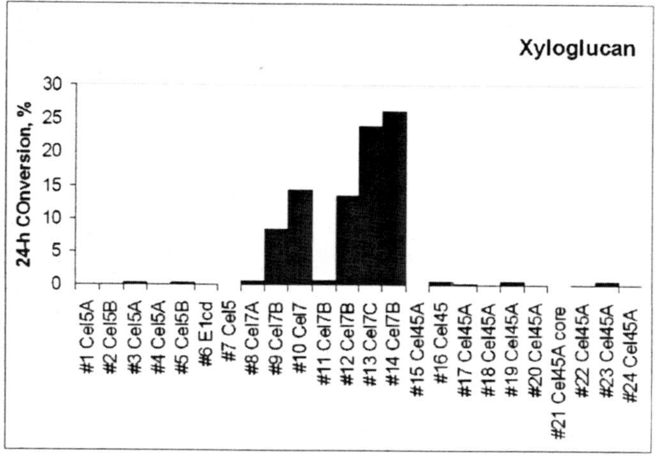

Figure 2. P-nitrophenyl cellulose (PNPC) activity was determined using method of Deshpande et al. (15) modified to a 96 well microplate format, by spectrophotometric measurement of p-nitrophenol produced after 30-min hydrolysis of PNPC (2.5 mM) in 50 mM sodium acetate pH 5.0 at 40 °C. Prior to hydrolysis, the enzymes were diluted in 50 mM sodium acetate pH 5.0 to give less than 8% conversion. Mannan, Xyloglucan and PASC at 5 mg/ml were digested by 0.5 mg enzyme/g substrate (0.25 mg/g for PASC) for 21 hours at 50 °C, and resulting reducting sugars determined by PHBAH.

when mixed with *T. reesei* broth was capable of achieving the same amount of hydrolysis after 24 hours with half the total loading of broth compared to the unmixed broths from *T. reesei* or *Fungus Z* (Figure 3). Proteins in these broths were identified by proteomic methods, including 2-D gels and mass spectrometry. Protein broths were also fractionated by anion exchange chromatography followed by Phenyl Sepharose hydrophobic interaction chromatography (G.E.). These semi-purified protein fractions were mixed with Celluclast supplemented by BG, and their ability to boost PCS hydrolysis was measured (Figure 4). Identification of the boosting proteins in these fractions was obtained by MALDI mass spectroscopy after SDS PAGE analysis.

Proteins tentatively identified as capable of increasing PCS hydrolysis were cloned, and expressed in fungal broths as monocomponent proteins for testing in addition to the best current fungal protein broths. In this way combinatorial mixtures of monocomponent mixtures could be tested for synergy or enhancement in hydrolysis of PCS. Some proteins identified as boosting Celluclast activity had little or no activity alone on PCS. In each case the percent cellulose conversion of the new mixtures was compared at various enzyme loadings, allowing the determination of minimum required protein for a

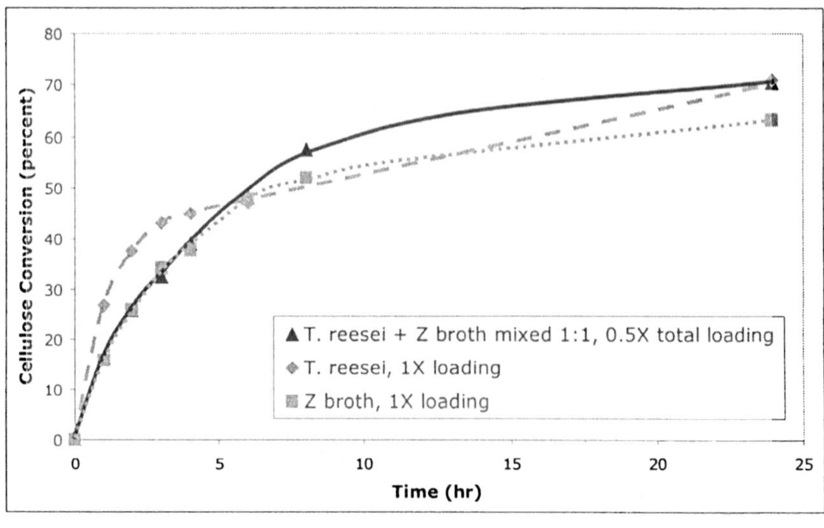

Figure 3. The mixture of T. reesei and Fungus Z broths gives synergetic activity in hydrolysis of PCS, requiring half the total loading to reach the same hydrolysis in 24 hours. Conditions are 50 mM acetate pH 5, 50 °C on washed PCS at 1%. Released sugars are measured by PHBAH. (Adapted with permission from reference 18. Copyright 2005 Bioscience and Industry.)

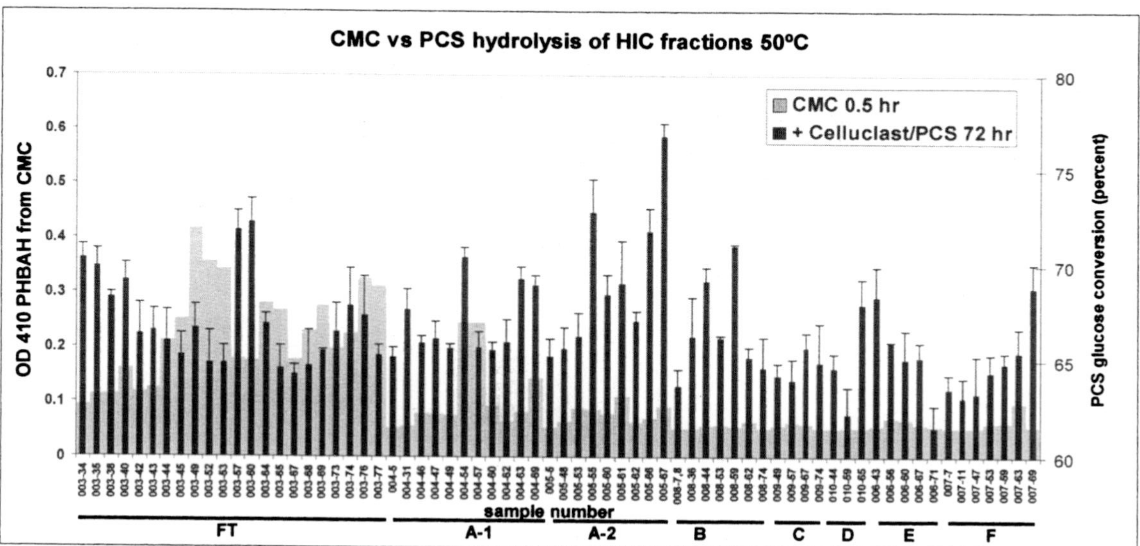

Figure 4. FPLC fractionation by anion exchange and HIC followed by measurement of hydrolysis of carboxymethyl cellulose (CMC) or enhancement of Celluclast activity on PCS. Conditions are 1% CMC or washed PCS in 50 mM citrate pH 5 at 50 °C. Released sugars are measured by PHBAH. Celluclast was included at a level that produced 62% conversion in this assay.

specific hydrolytic endpoint. An example of one addition with profound effect on hydrolysis activity of improved cellulase mixtures was members of the family 61 (Figure 5).

Additionally, a basal strain of *Trichoderma* with increased BG expression was mutagenized by treatment with N-methyl-N'-nitro-N-nitrosoguanidine to increase cellulolytic activity and used as a parent strain for further transformations.

Finally, the genes comprising the most effective cellulolytic mixtures were used to transform *T. reesei* in numerous combinations, producing hundreds of strains. These were tested for expression of individual activities and hydrolysis of PCS. Several similar strains surpassed all others in production of cellulolytic activity on PCS, yielding enzymatic improvements of about 6-fold over benchmark activity (Figure 6).

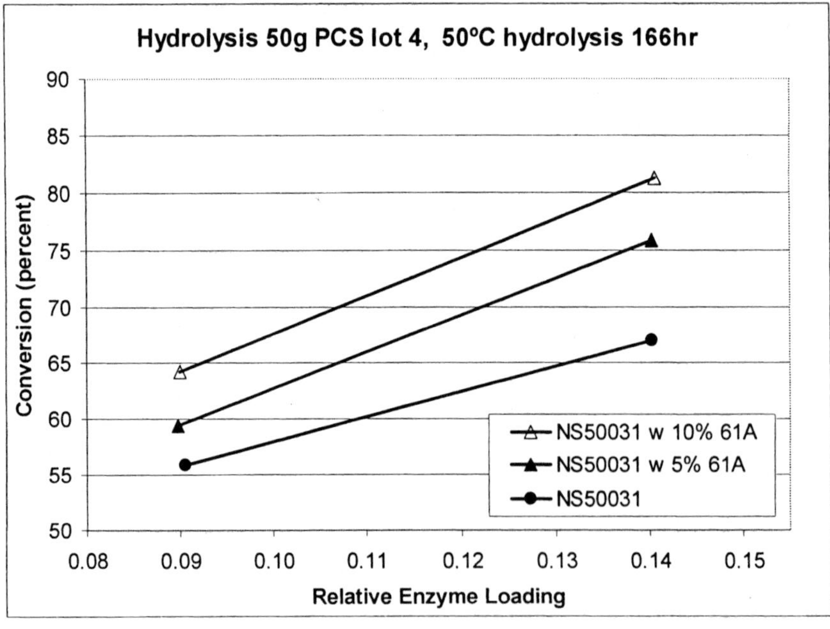

Figure 5. The cellulolytic activity of an improved enzyme mixture (NS50031 is a pilot-scale production run for customer sampling) is enhanced by the replacement of increasing amounts of protein by a member of the 61 family. NS50031 is used in a 50 g assay with 6.7% cellulose at 50 °C with replacement of 0, 5, or 10% of protein. Data represents the average of two flasks. The line connecting the data points is for graphical clarity only.

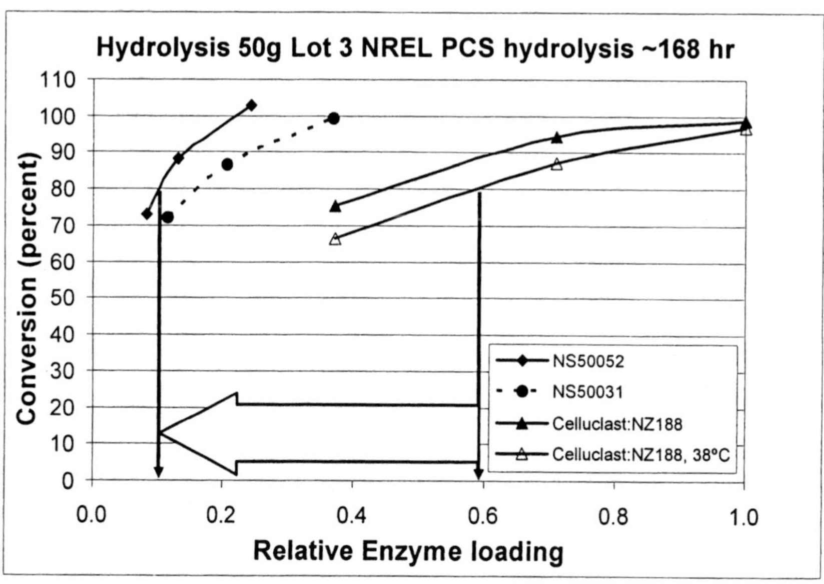

Figure 6. Improvement in enzymatic PCS conversion with pilot-scale production broths is 6-fold improved over initial benchmark enzyme conditions (1% Novozym 188 in Celluclast 1.5 L at 38 °C). NS50052 and NS50031 are pilot-scale production runs for customer sampling.

Using a multifaceted and integrated approach, we incorporated improvements to single enzymes, additions to the synergistic suite of enzymes, and modifications to the expression organism *T. reesei*. These improvements, combined with the improvements in pretreatment and production economies have led to a 30-fold reduction in the cost of PCS degrading enzymes, here efficiently produced in a single organism. In 2001 the cost of enzymes estimated by the NREL metric (*19*) (with assumptions of crop harvest yields, transportation costs, ethanol selling price, and scaling from laboratory results to the Nth stage plant with start-up in 2010) to produce ethanol from corn stover was over $5/gallon. This jointly funded collaboration between U.S. Department of Energy and Novozymes reduced this cost to $0.10 to $0.18 (*20*). Further reduction in enzyme costs appear possible, but the next steps will be the development of scaled processes utilizing these enzymes in collaboration with bioethanol producers, and expansion of this technology to other feedstocks. Customization of the enzyme mixtures is also expected to be necessary for other feedstocks and pretreatment methods.

References

1. Perlack, R.; Wright, L.; Turhollow, A.; Graham, R.; Stockes, B.; Erback, D. In *Biomass as Feedstock for a Bioenergy and Bioproducts Industry: The Technical Feasibility of a Billion-ton Annual Supply*, April **2005**, http://www1.eere.energy.gov/biomass/pdfs/final_billionton_vision_report2.
2. Bourne, Y.; Henrissat, B. *Curr. Opin. Struct. Biol.* **2001**, *11(5)*, 593-600.
3. Kubicek, C. P.; Eveleigh, D. E.; Esterbauer, H.; Steiner, W.; Kubicek-Pranz, E. M. In *Trichoderma reesei Cellulases*, Royal Society of Chemistry: Cambridge, UK., **1990**.
4. Himmel, M. E.; Adney, W. S.; Baker, J. O.; Elander, R.; McMillan, J. D.; Nieves, R. A.; Sheehan, J. J.; Thomas, S. R.; Vinzant, T. B.; Zhang, M. In *Fuels and Chemicals from Biomass*; Saha, B. C.; Woodward, J.: Eds., American Chemical Society: Washington, DC, **1997**, pp 2-45.
5. Cherry, J. R. In *Cellulase cost reduction*, nternational Energy Agency-Task 39 Bioenergy-Ethanol Subtask meeting, Copenhagen, Denmark, **2003**.
6. Hsu, T. A. In *Handbook on Bioethanol: Production and Utilization*;, Wyman, C.: Ed., Taylor and Francis: Washington, DC, **1996** , p 179-212.
7. von Sievers, M.; Zacchi, G. *Bioresource Technology*, **1996**, *56*,131-140.
8. Lynd, L. R.; Weimer, P. J.; Zyl, W. H-van; Pretnoius, I. S. *Microbiol. Mol. Biol. Rev.,* **2002**, *66*, 506-577.
9. Schell, D. J.; Farmer, J.; Newman, M.; McMillan, J. D. *Appl. Biochem. Biotechnol.*, **2003**, *105(108)*, 69-85.
10. *Biomass Publications of the U.S. Department of Energy Biomass Program*, http://www1.eere.energy.gov/biomass/publications.html, **2006**.
11. *Standard Biomass Analytical Procedures/NREL Laboratory Analytical Procedures of the U.S. Department of Energy Biomass Program,* http://www1.eere.energy.gov/biomass/analytical_procedures.html. **2006**.
12. Lever, M. *Biochem Med.* **1973**, *7(2)*, 274-81
13. Vlasenko, E.; Xu, F.; Cherry, J. R. *Proc. 25th Symposium on Biotechnology for Fuels and Chemicals*, Breckenridge, CO, **2003**.
14. Fidantsef, A. L.; Gorre-Clancy, B.; Yi, J. H.; Teter, S.A.; Lamsa, A. M.; Lamsa, M. H.; Cherry, J. R. *Proc. 25th Symposium on Biotechnology for Fuels and Chemicals*, Breckenridge, CO, **2003**.
15. Deshpande, M. V.; Eriksson, K.-E.; Pettersson, L.G. *Anal. Biochem.* **1984** *138*, 481-487.
16. Eisen, M. B.; Brown, P. O., *Methods Enzymol.* **1999**, *303*, 179-205.
17. Bashkirova, E. V.; Berka, R. M. *Proc. 25th Symposium on Biotechnology for Fuels and Chemicals,* Breckenridge, CO, **2003**.
18. Takagi, S.; Sakaguchi, H; Xu F. *Bioscience and Industry* **2005**, *63(7)*, 451-456.

19. *Lignocellulosic Biomass to Ethanol Process Design and Economics Utilizing Co-Current Dilute Acid Prehydrolysis and Enzymatic Hydrolysis for Corn Stover*, http://www.nrel.gov/docs/fy02osti/32438.pdf, **2002**.
20. *Novozymes and NREL Cut Cost of Converting Biomass to Ethanol*, http://www.eere.energy.gov/news_detail.cfm?nes_id=9004, **2005**.

Chapter 3

Biocatalytic Conversion of Granular Starch to Industrial Chemicals

Manoj Kumar, Jeff Pucci, Jay Shetty, Gopal Chotani, and Karl Sanford

Genencor International, A Danisco Company, 925 Page Mill Road, Palo Alto, CA 94304

Biocatalytic systems to economically produce biochemicals using less expensive feedstocks like granular starches are an attractive alternative to petroleum based chemical synthesis. This is due, in part, to significant advances in our understanding and application of genetic and metabolic engineering of host organisms and granular starch hydrolyzing enzymes. This book chapter explores a GSHE (granular starch hydrolyzing enzyme) process for the conversion of granular starch feedstocks to glucose, and the concomitant conversion of glucose by microorganisms and/or enzymes, to desired chemical products. In this process, conversion of glucose to products parallels its formation by enzymatic catalysis. This reduces enzyme inhibition and speeds the conversion of granular starch to products. Furthermore, such a process provides key glucose-controlled biocatalytic conditions, which are critical in many fermentative processes.

Introduction

Industrial fermentations are used for the manufacturing of bioproducts (*1*). They predominantly use glucose as a feedstock for the production of proteins, enzymes, and chemicals. These fermentations can be batch, fed-batch, or continuous, operate under controlled-substrate feeding, and conditions that form minimal byproducts. Substrate-controlled feeding, pH, dissolved oxygen, and medium composition are critical operating conditions that must be controlled during fermentation, to optimize fermentation time, yield, and efficiency. Currently used methods and feedstocks, however, have drawbacks that reduce the efficiency of the fermentation processes.

Traditionally cooked grains and starchy feedstocks with thermostable enzymes are used to begin the process of converting soluble starch to fermentable sugars, mainly glucose, although the conversion employs an energy intensive jet cooking process. Glucose is useful as a starting substrate in a multitude of chemical and biological synthetic applications, but its use for fermentation processes is disadvantageous for several reasons. Glucose syrups of purity levels greater than 90% are relatively expensive. In addition, the presence of even moderate glucose concentrations increases the susceptibility of the fermentation system to microbial contamination, thereby resulting in an adverse effect upon the production efficiency. Moreover, even the presence of low to moderate levels of glucose in the fermentation vat adversely affects its conversion to the desired end product, due to enzymatic inhibition and/or catabolite repression (*2*). Various attempts have been made to reduce the costs of industrial fermentation by utilizing less expensive substrates than glucose (*3*). Despite numerous approaches there remains a need to develop methods to more efficiently produce desired chemicals and biochemicals from economical, non-glucose substrates (*4*).

Renewable feedstocks such as grain cereals and starchy tubers are comprised of two major polysaccharides: amylose (80-70%) and amylopectin (20-30%). These feedstocks are resistant to degradation and need a jet cooking pretreatment to allow facile enzymatic digestibility. Amylase enzymes responsible for the hydrolysis of starch fractions of cereals to glucose perform sub-optimally, mainly due to reversion reactions. Therefore, keeping solublized starch available to amylases, and the product concentration at the highest levels, can help maintain the stability and activity of amylase enzymes. Granular Starch Hydrolyzing Enzymes (GSHEs) that can hydrolyze granular (uncooked) starch have been reported as early as 1944 (*5*). Subsequently, there have been many research publications on the characterization and production of these GSHEs. While there has been considerable interest in granular starch hydrolysis, commercial use of these enzymes has been limited to Sake brewing in Japan. This is because of the high cost of commercial GSHEs and their sub-optimal solid-state tray fermentations. Genencor International has successfully developed, a GSHE technology to make granular starch hydrolysis attractive to produce chemicals such as ethanol (*6*) through high expression submerged culture fermentation.

Genencor International's research (*7*) on continuous biocatalytic systems using sequential enzyme reactions for directly processing the granular starch component of cereals and tubers to biochemicals addresses, issues related to (i) improved productivity, (ii) enzyme inactivation, (iii) reduced energy consumption, (iv) higher product yields, (v) mass transfer limitations, and (vi) saving on capital expenses; and thus overcomes key existing limitations for granular starch conversion to industrial chemicals (*8*). Advantages of Genencor's biocatalytic conversion of granular starch (*7*) to value-added chemicals include, (a) commercial viability, simplicity, and better economic feasibility; (b) prevention of reversion reactions by concurrent conversion to bioproducts; (c) feasibility of quantitative conversion, (d) elimination of byproducts; (e) higher productivity and yield on carbon; (f) production capacity enhancement; and (g) most importantly, reduced energy consumption. This biocatalytic conversion concept is novel because a multi-enzyme process for converting renewable starch based substrates to value-added commercial ingredients has not yet been commercially demonstrated.

Various industrial chemicals such as organic acids (gluconic, citric, succinic, acetic, ascorbic, lactic, 3-hydroxy propionic acid), solvents (acetone, ethanol, glycerol, butanol, 1,3-propanediol), amino acids (glutamate, lysine, methionine), antibiotics (penicillin), and industrial enzymes can be made from fermentable sugars derived from granular starch using Genencor's biocatalytic systems. In this book chapter, we illustrate a case study of gluconic acid production from granular starch using a biocatalytic system, and compare it with a current manufacturing process (*9-10*). We provide data to support the use of this method for conversion of cereal and tuber-derived feedstocks to other industrially relevant biochemicals.

In-Vitro Biocatalytic Conversion of Granular Starch to Gluconic Acid

Today, gluconic acid is one of the major industrial biochemicals produced using fermentation technology (*9*). Both, sodium gluconate and gluconic acid are used in various applications, such as in dairy, food, pharmaceutical, cleaning, textiles, cement, and metallurgy. The gluconate and glucono-d-lactone markets are substantial (Table I) and are growing rapidly. Key producers of gluconic acid, at present, include Purac, Fujisawa, Akzo, Roquette, Glucona, and Jungbunzlauer.

Table I. Gluconic Acid and Gluconate Market Data

Chemical	*Volume (M T/yr)*	*Price ($$/Kg)*	*Market (MM$/yr)*
Sodium Gluconate	60,000	1.50	90
Glucono-d-lactone	15,000	3.50	53

SOURCE: Chemical Market Reporter, 1998, July 20

Current Gluconic Acid Processes

Several processes for the production of gluconic acid by chemical and biochemical means are reported in the literature (*10,12*). Chemically, gluconic acid is produced by oxidation of the aldehyde group in glucose, and the subsequent reaction of the acid with an appropriate base to make gluconate. The current biochemical manufacturing process for gluconic acid is by *Aspergillus niger* fermentation of glucose (*9*). The formation of gluconic acid by this process is described in the following equation:

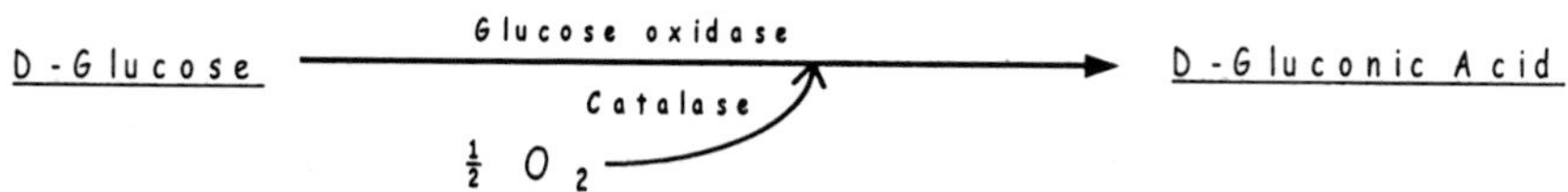

The *Aspergillus* process requires a high concentration of glucose (>25 wt %), high aeration rates, and dissolved oxygen, the latter is achieved by applying up to four bar of air pressure. This viable whole cell process can have a maximum of 30% solids in sugar content, as a further increase in sugar inhibits cellular metabolism. An alternative to this whole cell process is a patented (*11*), but not practiced, technology of using enzymes and glucose. However, this method is not competitive with the whole cell process due to its economic need of using 30-60% glucose concentrations, at which glucose is inhibitory to enzymes. Using a dilute glucose feed would yield a dilute gluconic acid product stream, which would be economically non-viable for the recovery of solid gluconate. Thus, this patented (*11*) enzymatic process not only requires high enzyme dosage but also results in high viscosity of the reaction medium due to increased sugar concentration, thereby reducing the rate of oxygen transfer.

New Gluconic Acid Process

The granular starch based gluconic acid production process (*7*) can successfully compete with the current *A. niger* gluconic acid technology for several reasons:

1) This process needs significantly lower dosage of enzymes.
2) The steady state concentration of glucose is low in the bioreactor, resulting in reduced substrate and/or product-based inhibition of enzymes.
3) The cost of granular materials is significantly less; granular starch versus D-glucose.
4) There is control over feed stock concentration and thus oxygen transfer limitations do not exist.
5) Better carbon efficiency, as putting *A. niger* cell mass in place in the current process uses 20% of total glucose.
6) Absence of waste disposal issues.
7) A simpler down-stream recovery process.

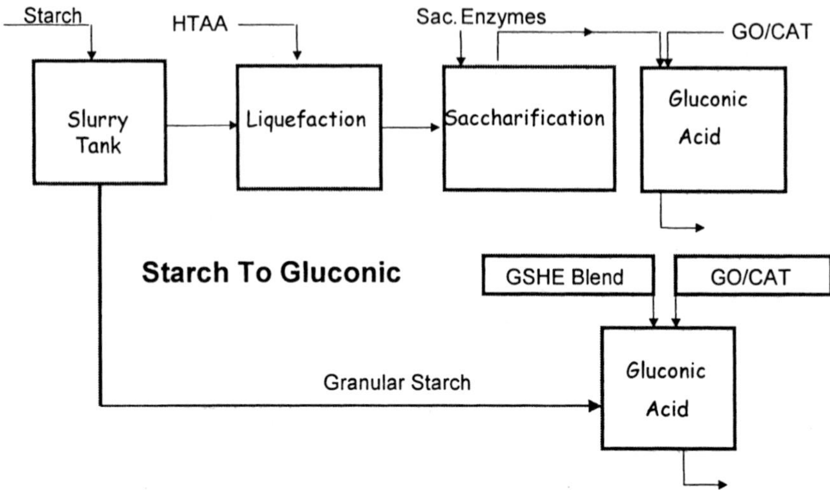

Figure 1. Single step GSHE bioconversion of granular starch to gluconic acid versus the conventional process. HTAA, high temperature α-amylase; GO, glucose oxidase; CAT-catalase.

The key advantages of this simple and economical *in-vitro* biocatalytic process for granular starch conversion to gluconate are: (a) reduced energy consumption, (b) prevention of product inhibition of amylase enzymes by concurrent conversion to gluconate, (c) feasibility of quantitative conversion, (d) elimination of byproduct formation, (e) higher productivity, (f) increased production capacity, and (g) higher yield on carbon.

Experimental Methods

Enzymes used in this work were assayed for activity using methods adopted from Bergmeyer and Jaworek (*13*) in Methods of Enzymatic Analysis. All products and substrates were measured in the reaction using HPLC, and quantitation of products was referenced to a concentration based calibration plot that was generated. Glucose concentration was assayed using both HPLC and Monarch glucose analyzer. Before analysis, samples for the enzymatic reaction were appropriately diluted (10-100 fold) to meet the linear range of the calibration curve generated. All the experiments were performed in 1 L laboratory fermenters (Applicon) equipped with an air supply system, stirring, pH control, temperature control, foam control, and dissolved oxygen (DO) probe. Unless otherwise mentioned, all the experiments were underaken using the following parameters: temperature 45 °C, pH 5 (buffered using 50 mM citrate), agitation between 600-900 rpm to maintain dissolved oxygen above 30% saturation, working volume 300 mL, and pH control using 10% sodium hydroxide solution. In the calculation of the conversion, the dilution effect of adding sodium hydroxide solution was taken into account. Enzyme activities of glucose oxidase, catalase are expressed in Titrimetric Units and Baker Units. One Titrimetric unit will oxidize 3.0 mg of glucose to gluconic acid in 15 min under assay conditions of 35 °C at pH 5.1. One Baker unit of catalase decomposes 264 mg of hydrogen peroxide in 1 h under assay conditions of 25 °C at pH 7. One Unit of GSHE liberates one micromole of reducing sugar (expressed as glucose equivalent) in 1 min under assay conditions of 50 °C at pH 5. Cornstarch slurry (30%) in 10 mM acetate buffer pH 5.0 was prepared and brought to 40 °C, 1100 rpm, and 118 DO. GSHE (250 mg), 150 μL of Distillase® L-400 (350 AU/g; sp 1.15), 1250 μL of Oxygo®, and 1500 μL of Fermcolase® were mixed. This led to an initial gluconate production rate of 25 g/L/h. The degree of hydrolysis of starch to glucose, and glucose to gluconate, was measured over the course of reaction (Figure 2).

Results

Preliminary experiments (*7*) illustrated an initial good conversion rate of granular starch derived from corn to gluconate using this enzymatic system. Glucose was produced and concomitantly converted to the final product thus preventing its accumulation. This method thus provides an attractive alternative for production of gluconate from granular starch when compared to direct conversion of glucose to gluconic acid using glucose oxidase and catalase at the same enzyme dosage level. All the enzymes remained stable during this biocatalytic process. A comparison between two granular starches (wheat versus corn) was made and found that conversion of wheat granular starch to gluconate was superior to that of corn granular starch (Figure 3).

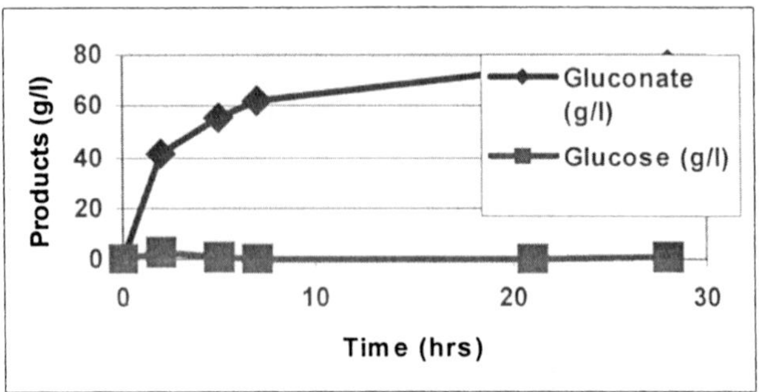

Figure 2. In vitro biocatalytic conversion of granular corn starch to gluconic acid with GSHE, OxygO, Distillase and Fermcolase enzymes.

As shown in Figure 3, wheat starch can also be efficiently converted to gluconate using Oxygo®, Fermcolase®, Distillase®, and Sumizyme (CU CONC™ granular starch hydrolyzing enzymes, 187 glucoamylase units/g of powder, Shin Nihon, Japan). Indeed, the results indicate that granular wheat starch is more amenable to bioconversion than granular corn starch when compared for the similar bioconversion time.

One of the key factors for successful commercial implementation of this technology, is to ensure that all the enzymes needed for granular starch conversion to gluconate remain stable. Hydrogen peroxide, produced as a byproduct in the conversion of glucose to gluconate, is an oxidative irreversible inactivator of biocatalysts. In the biocatalytic system, since the glucose concentration in the reaction system is effectively zero, hydrogen peroxide buildup does not occur. It is thus possible to take these bioconversion reactions to theoretically maximal yields by preventing inactivation of enzymes by H_2O_2.

Fermentation of Granular Starch to Industrial Chemicals

A biocatalytic system for converting granular starch to industrial chemicals is not only applicable to enzymatic conversions but also to fermentative conversion using granular starch (*14*). We report here three examples of fermentative conversion of granular starch to chemicals namely 1,3 propanediol, lactic acid, and succinic acid.

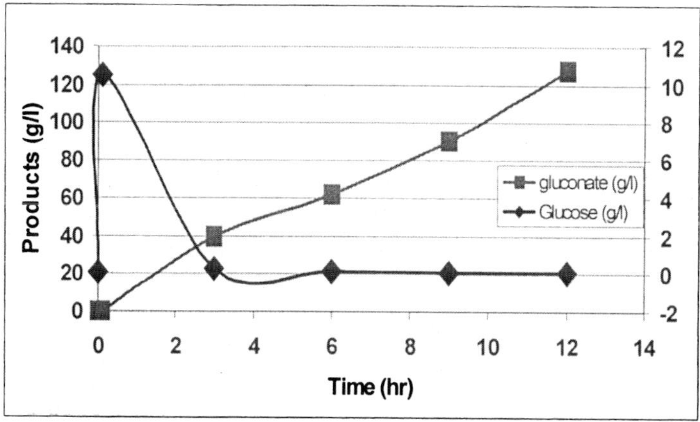

Figure 3. In vitro biocatalytic conversion of granular wheat starch to gluconic acid with Sumizyme, OxygO, Distillase and Fermcolase enzymes.

Experimental Methods

Experiments to convert granular starch to 1,3-propanediol (*15*) and succinic acid were performed at 34 °C and pH 6.7, and for conversion to lactic acid at 34 °C and pH 6.4. Granular starch (Corn) was used in a slurry form following pasteurization at 65 °C for 30 min. Desired enzymes and requirements specific for 1,3-propanediol production (20 mg spectinomycin and 1 mg vitamin B_{12}) were added as 0.2 μm filtered solutions in DI water. TM2 fermentation medium consisted of the following components: Potassium dihydrogen phosphate 13.6 g/L, dipotassium hydrogen phosphate 13.6 g/L, magnesium sulfate hexahydrate 2 g/L, citric acid monohydrate 2 g/L, ferric ammonium citrate 0.3 g/L, ammonium sulfate 3.2 g/L, and yeast extract 5 g/L.

Results

For control experiments 30% granular starch slurry (pasteurized) was made in a 2 L flask by combining with 200 mL of TM2 medium. Following the addition of the test enzymes (30 mL Ultra Filter concentrate of fermenter supernatant from a *Humicola grisea* fermentation run showing granular starch hydrolyzing activity (glucoamylase) and 0.4 mL of SPEZYME® FRED α-amylase liquid concentrate (Genencor), and 20 mg spectinomycin and 1 mg vitamin B_{12}, samples were taken from the vessel at varying reaction times, centrifuged, and the supernatants frozen to terminate enzyme action. The supernatants were subjected to HPLC analysis. In this experiment, hydrolysis of granular starch was monitored by measuring glucose formation. It was determined that 32.09g/L glucose accumulated in 3 h. Conversion of granular starch to glucose was demonstrated at a rate of 10 g/L/h at 34 °C and pH 6.7 (data not shown). Similarly, using 250 mg of Sumizyme resulted in an initial 16 g/L/h conversion of starch to glucose at pH 5.0 and 45 °C. These results indicate that both granular starch hydrolyzing enzymes have excellent kinetics for starch to sugar conversion.

For measurement of 1, 3-propanediol production, granular starch slurry was prepared in a minimal fermentation medium in a 1 L bioreactor. The pH of the slurry/broth was adjusted to 6.7 and controlled at 6.65 with NH_4OH. Then, the desired enzymes (30 mL ultra filter concentrate of fermenter supernatant of a *Humicola grisea* run showing granular starch hydrolyzing activity (glucoamylase), and 0.4 mL of SPEZYME® FRED liquid concentrate [Genencor] having α-amylase activity), and requirements specific for 1,3-propanediol production (30 mg spectinomycin and 2 mg vitamin B_{12}) were added as 0.2 μm filtered solutions in DI water. An inoculum of 1,3-propanediol-producing *E. coli* strain TTaldABml/p109F1 (Genencor International internal culture collection) taken from a frozen vial, was prepared in soytone-yeast extract-glucose medium. After the inoculum grew to OD 0.6, measured at 550 nm, two 600 mL flasks were centrifuged and the contents resuspended in 70 mL supernatant to transfer the cell pellet (70 mL of OD 3.1 material) to the bioreactor. Samples taken from the reaction vessel at varying times were centrifuged and the supernatants frozen to terminate the enzyme action. The supernatants were subjected to HPLC analysis. In this experiment, fermentation of granular starch to 1,3-propanediol was monitored by measuring glucose formation and its conversion to glycerol (an intermediate in the 1,3-propanediol pathway), and then to 1,3-propanediol. In 23.5 h, accumulation of glycerol and 1,3-propanediol amounted to 7.27 and 41.93 g/L, respectively (Figure 4). Conversion of granular starch to glycerol and 1,3-propanediol at 1.75 g/L/h rate was demonstrated for fermentative bioconversion of granular starch to 1,3-propanediol at 34 °C and pH 6.7.

To further assess the utility of the biocatalytic system, granular starch conversion to lactic acid (*16*) and succinic acid (*17*) was also examined. Granular starch fermentation to lactic acid was monitored by measuring glucose formation from granular starch using Sumizyme and its subsequent conversion to lactate using the lactate producing strain *Lactobacillus casei.* Pasteurized granular starch in a slurry form was mixed in *Lactobacilli* MRS medium (Difco), and 0.4 units of the enzyme were added to it. An inoculum of the lactate producing strain of *Lactobacillus casei* (ATCC 393 grown at 34 °C with nitrogen sparge at 0.6 slpm was added (46 mL of OD 24.2 measured at 550 nm) to the bioreactor. Samples taken from the reaction vessel at various times were

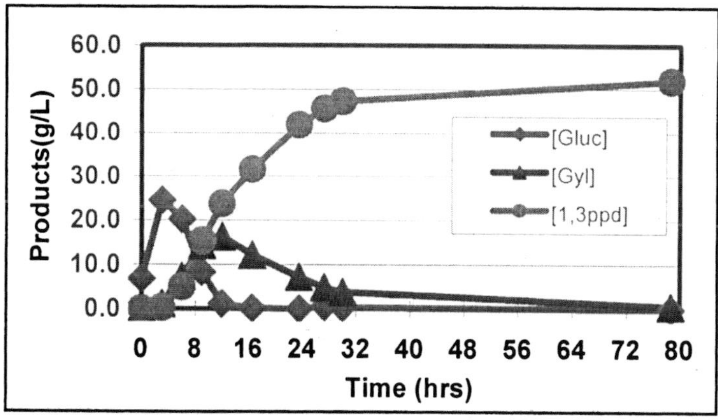

Figure 4. Biocatalytic conversion and concomitant fermentation of granular starch to 1, 3-propanediol

centrifuged, and supernatants were frozen to terminate the enzyme action. The supernatants were subjected to HPLC analysis. In this experiment, bio-conversion of granular starch was monitored by measuring glucose formation and its conversion to lactate. In 16.3 h, accumulation of lactate amounted to 61.75 g/L (Figure 5). In addition, the bioconversion of granular starch to lactate was demonstrated to be at a level of 3.79 g/L/h rate, at a temperature of 34 °C, and at pH 6.4 (Figure 5).

Fermentative conversion of granular starch to succinic acid was carried out in a 1 L bioreactor by measuring glucose formation from granular starch and its subsequent conversion to succinate, using 0.6 g of Sumizyme and a succinate producing *E. coli* strain, 36 1.6 ppc (Genencor International internal culture collection). Granular starch slurry was prepared in the minimal fermentation medium and, following enzyme addition, 80 mL of OD 14.3 measuring inoculum of succinate-producing strain 36 1.6ppc *E. coli* grown in TM2 medium from a frozen vial under nitrogen sparge at 0.6 slpm was added to the bioreactor. Samples from the reaction vessel, taken at various times, were centrifuged and the supernatants frozen. The supernatant was subjected to HPLC analysis. In this experiment fermentative conversion of granular starch to succinate was monitored by measuring glucose formation and its conversion to succinate (Figure 4). In 22 h, accumulation of succinate amounted to 26.5 g/L (Figure 6). The conversion of granular starch to succinate at 0.034 g/L/h rate was demonstrated for fermentative bioconversion of granular starch to succinate at 34 °C and pH 6.7.

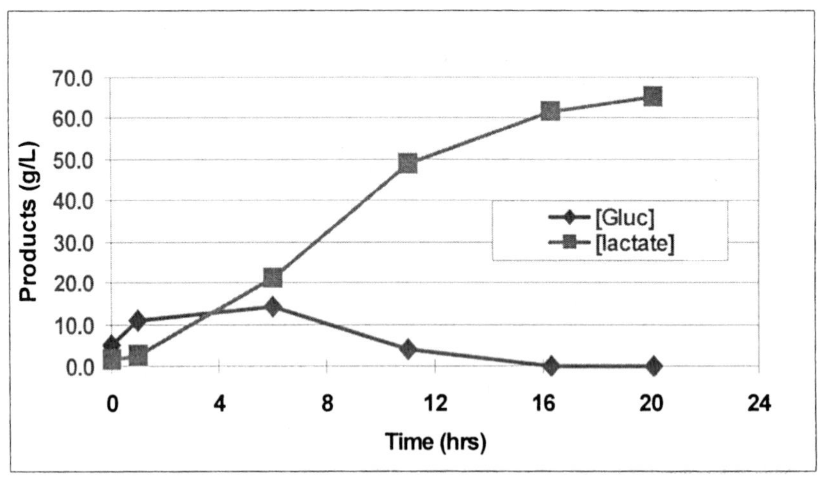

Figure 5. Biocatalytic conversion and concomitant fermentation of granular starch to lactic acid

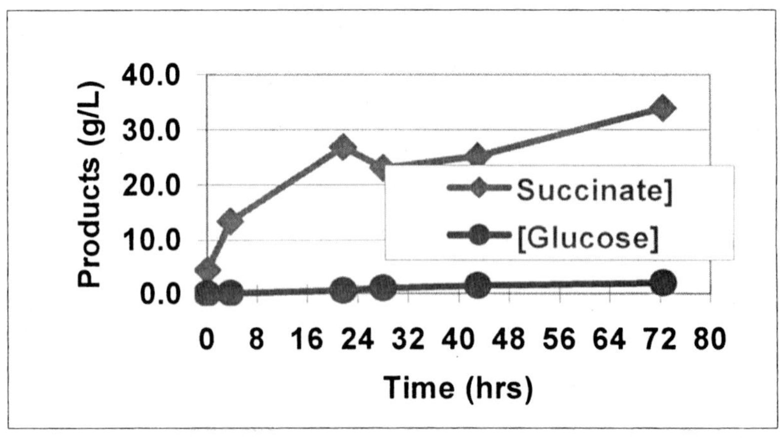

Figure 6. Biocatalytic conversion and fermentation of granular starch to succinic acid

Discussion

Hydrolysis of granular starch from grain and tubers to glucose occurs by the breakdown of α-1,4 and α-1,6 glycosidic linkages between the glucose units of starch. Conventional hydrolysis of starch to glucose is a three-step process. First, generally 30-35 wt % of granular starch slurry is cooked at 90-150 °C in the presence of thermostable α-amylase, then held at 90 °C for 1-3 h and cooled to 60 °C and treated with glucoamylase. As evident, this is an energy intensive (*18*) process. A process that can eliminate the energy demand, such as the use of granular starch hydrolyzing enzymes, is attractive. It was calculated (*4*) that the energy demand for conventional hydrolysis of starch if converted to ethanol, is about 10-20% of ethanol fuel value. Furthermore, liquefied starch produced through the conventional process has a several fold higher viscosity than the granular starch slurry and thus presents mechanical difficulties in handling. High viscosity glucose syrup made through conventional process when used in subsequent fermentations and bioconversions, also presents challenges in substrate and product inhibition. Finally, byproducts from Maillard reactions are an unwanted consequence of the conventional starch hydrolysis process, reducing overall efficiency and yield (*19*).

Biocatalytic conversion of granular starch to industrial products has thus attracted attention. Recently grain processors (*20*) have begun to employ the GSHE technology for producing bioethanol. However, the desired dose of GSHE enzymes for economical bioconversion, less than desired yield, possible enzyme inhibition by products, and risk of contamination (*21*) are some of the parameters that need to be optimized for a successful shift from conventional processing of starch to GSHE technology. The continuous bioconversion approach described in this work offers possible solutions for minimizing or eliminating the contamination issues.

Starch hydrolyzing enzymes are present widely in nature (*22*). Both cereal and tubers have granular starch hydrolyzing activities, as these enzymes are essential for seed and tuber germination. Many grains express multiple granular starch hydrolyzing enzymes during seed germination process (*23*). Thus screening for the most promising granular starch hydrolyzing enzymes (*24*), identification of corresponding genes, their efficient and economical expression in suitable expression hosts by fermentation, will be essential for commercial success and to the benefits of the GSHE technology. Furthermore, granular starch hydrolyzing enzymes will need to have desired physical and chemical stability for the reaction conditions they would be used in, excellent kinetic parameters for hydrolysis of granular starch including resistance to inhibition. Last but not least, for reduced digestion time and lower temperature fermentations, synergism using a multiple enzyme cocktail may also be needed for the efficient hydrolysis of granular starch.

Summary

This chapter describes the bioconversion of renewable feedstocks to industrial chemicals. A technology, developed by Genencor International, to harness granular starch as carbon feedstocks for conversion to industrial products, and to make available its bioengineered enzymes to convert granular starch into fermentable sugars, has been illustrated. This technology provides a means for the production of desired bioproducts by enzymatic conversion of renewable bio-based feedstock substrates. The concept of using granular starch for manufacturing industrial chemicals has several incentives that can be explored and implemented.

Acknowledgements

Authors wish to acknowledge Dr. Roopa Ghirnikar for editorial help in preparing this manuscript.

References

1. Chotani, G.; Dodge, T.; Hsu, A.; Kumar, M.; LaDuca, R.; Trimbur, D.; Weyler, W.; Sanford, K. *Biochemica Biophysica Acta*, **2000**, *1543*, 434.
2. Dien. B. S.; Nicols, N. N.; Bothast, R. J. *J. Ind. Microbiol. Biotechnol.* **2002**, *29*, 221.
3. Linko, P. *Biotechnol Adv.* **1985**, *3*:39.
4. Robertson, G. H.; Wong, D. W.; Lee, C. C.; Wagschal, K.; Smith, M. R.; Orts, W. J. *J Agric Food Chem.* **2006**, *54*, 353.
5. Balls, A. K.; Schwimmer, S. *J. Biol. Chem.* **1944**, *156*, 203.
6. Shetty, J. K.; Lantero, O. J.; Dunn-Coleman, N. *Internat. Sugar J.* **2005**, *107*, 1281.
7. Chotani, G. K.; Kumar, M.; Pucci, J.; Shetty, J.K.; Sanford, K. J. **2003**, *Methods for Producing End Products From a Carbon Substrate.* US03/03532, WO 03/066816.
8. Schugerl, K. *Adv Biochem Eng Biotechnol.* **2000**, *70*, 41.
9. Fong, W. S. *Fermentation Processes: Process Economic Program*, Stanford Research Institute, **1975**; Report No. 95.
10. Milsom, P. E. and Meers, J. L. *Comprehensive Biotechnology* **1985**, *3*, 681-702.
11. Vroemen, A. J. and Beverini, M. Enzymatic Poduction of Guconic Aid or Its Salts, **2000**, US00/5897995.

12. Anastassiadis, S.; Aivasidis, A.; Wandrey, C. *Appl Microbiol Biotechnol.* **2003**, *61*, 110.
13. Bergmeyer, H. U.; Jaworek, D. In Methods of Enzymatic Analysis, Bergmeyer, H. U.; Ed.; Third edition, Verlag Chemie: Weinhem, **1983**.
14. Sanford, K.; Soucaille, P.; Whited, G.; Chotani, G. K. *Current Opinion in Microbiology* **2002**, *5*, 318-322.
15. Zeng, A. P.; Biebl, H. *Adv. Biochem. Eng. Biotechnol.* **2002**, *74*, 239-262.
16. Lawford, H. G.; Rousseau, J. D. *Appl Biochem Biotechnol.* **2002**, *98-100*, 429-448.
17. Barnes, S. P.; Keller, J. *Water Sci Technol.* **2003**, *48*, 155-62
18. Kelsall, D. R.; Lyons, T. P. In *The Alcohol Text Book*; Jaques, K.; Lyons , T. P.; Kelsall, D.R.; Eds.; Nottingham University Press: Nottingham **1999**, pp 7-24.
19. Galvez, A. *Ethanol Producer* **2005**, *11*, 58-60.
20. Berven D., The making of Broin Project X, *Ethanol Producer* **2005**, *11*, 67-71.
21. Hattori, A.; Miura, M.; Takahashi, M.; Uchida, N.; Furuya, K.; Hosoya, T. Isolated cultures of *Pestalotiopsis funere,* **1997**, US Patent 5604128.
22. Greenwood, C. T.; Milne, E. A. *Adv Carbohydrate Chem.* **1968**, *23*, 281-286.
23. Kossman, J. and Lloyd, J. *Crit Rev. Biochem. Mol. Biol.* **2000**, *35*, 141-196.
24. Dunn-Coleman, N.; Fiske, S. M.; Lantz, S. E.; Neefe-Kruithof, P.; Pepsin, M. J.; Shetty, J. K. *Acid-stable alpha amylases having granular starch hydrolyzing activity and enzyme compositions*, **2006**, US 20060003408A1.

Advances in the Application of Industrial Enzymes on Carbohydrate Food Materials

Chapter 4

New Enzymes and Products of the Anhydrofructose Pathway of Starch Catabolism

Shukun Yu, Karsten Kragh, and Andrew Morgan

Danisco Innovation, Danisco A/S, Langebrogade 1, DK 1001, Copenhagen, Denmark

As one of the leading industrial and food enzyme producers and suppliers, Danisco A/S has developed an α-1,4-glucan lyase (EC 4.2.2.13) from marine red algae *Gracilariales,* that produces 1,5-anhydro-D-fructose (AF) from starch. AF is a known versatile building block that can be used for the synthesis of an untold number of interesting compounds, either by enzymatic or chemical methods, or a combination of both. AF is another compound that places starch as a biorenewable material for the production of fine chemicals, that could have a future in replacing petroleum materials. These AF derived compounds have antioxidant, antimicrobial, anti-blood clotting, and anti-tumor activities. The technology for converting starch to AF and its valuable derivatives is known as *The Anhydrofructose Technology*. The enzymes found in red algae and fungi, that convert starch or glycogen to AF, and further convert AF to its various metabolite derivatives, have been referred to as *The Anhydrofructose Pathway.*

Introduction

For the sake of our environment and the limited supply of fossil fuels, it is imperative that methods and processes that utilize renewable materials, such as sucrose, starch, and celluloses, are developed. Furthermore, the chemical industry needs to become more eco-friendly or "green", by transforming conventional chemical processes to biocatalysis processes (bioprocesses). It is envisaged that our society will be more and more based on a bio-economy with sustainable growth. To reach these goals, scientists at the Genencor Division of Danisco A/S at Palo Alto (CA, USA) have developed processes for converting agricultural waste, namely, cellulose and hemicellulose products such as Corn Stover, to fermentable sugars that can then be used to produce bioethanol. At our division of Danisco Innovation (Danisco A/S) in Copenhagen (Denmark), we have focused on the bioconversion of starch to value-added products (*1-5*). To make such bioconversion possible, a great effort has been made to discover and identify new biocatalysts (*6-12*). For diversity screening, we turned our interest to marine red algae (*6-8*). In contrast to higher green plants, the starch in red algae is accumulated in the cytosol and not in an organelle (such as chloroplasts), similar to the accumulation of glycogen in fungi and aminal cells (*13*). It was, therefore, expected that red algae may possess unique machinery for metabolizing starch. As a result, we discovered a new α-glucan degradation pathway in red algae, namely, *The Anhydrofructose Pathway* (Figure 1) (*6-8*). Further research suggests that the anhydrofructose pathway may also occur in other organisms, such as bacteria (*12,14-15*), fungi (*16-18*) and mammals (*19-21*). Among the enzymes from this pathway that have potential industrial applications are α-1,4-glucan lyase that forms AF from starch, AF dehydratase that converts AF to microthecin (an anti-fungal compound), and ascopyrone tautomerase that forms ascopyrone P (APP) (Figure 1) which has antibacterial, antioxidant, and antitumor effects.

Beside the discovery of enzymes (*6-12*), research on the AF pathway in our laboratory has also covered gene cloning and expression (*22-23*), characterization and mechanistic studies (*24-29*), characterization of metabolites (*10-12,30-32*) and their function (*33-37*), and metabolic studies of these metabolites (*38-39*). With respect to the regulation of the AF pathway, our work has focused on the mechanisms at the transcription and enzyme levels (feed-back inhibition and covalent modification) (11,22-23). Together with research from other laboratories (*14,16-18,40-42*), we propose that the AF pathway is operative under stress conditions in red algae, fungi and possibly other organisms, and the metabolites of this pathway are used to counteract the stresses by acting as antioxidants and antimicrobials. In this book chapter, production of these metabolites by biocatalysis using enzyme reactor technology is discussed, as well as possible applications of these metabolites for food and non-food applications.

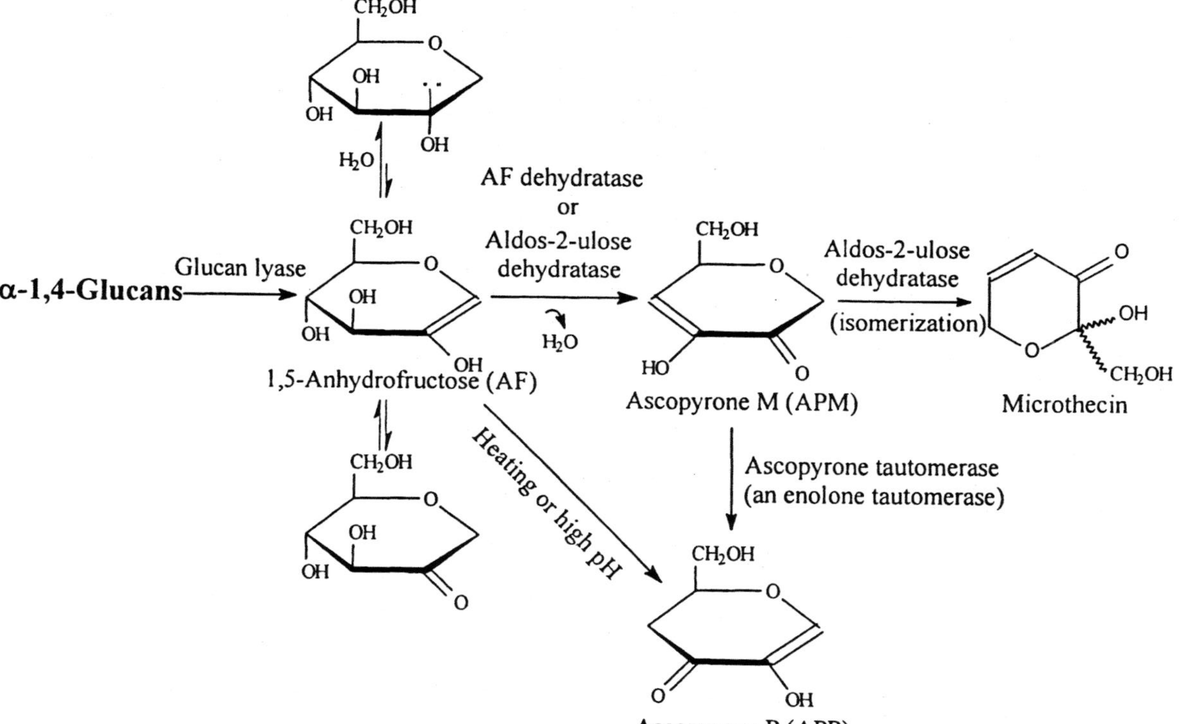

Figure 1. Enzymatic conversion of starch to microthecin and ascopyrone P (APP) via 1,5-anhydrofructose (AF), the Anhydrofructose Pathway. The formation of APP can also be achieved by non-enzymatic method as indicated.

Molecular and Enzymatic Features of Glucan Lyase Catalyzed Reactions

Glucan Lyase Sources and Their Basic Molecular Features

We isolated and studied six glucan lyases from subspecies of the red algae *Gracilariopsis lemaneiformis,* collected from Qingdao (Tsingtao, Shandong, China), Santa Cruz (California, USA), and the Araya Peninsula (Sucre, Venezuela), and *Gracilaria verrucosa* from the Araya Peninsula (Sucre, Venezuela) (*6-8,23-25*). Dr. Abe and Dr. Yoshinaga from Kagoshima University studied the lyase from the red alga *Gracilaria corda*, collected from the Kagoshima coast of Japan (*43*). The fungal lyases we studied were from *Mochella costata*, *Morchella vulgaris* of morels, *Peziza ostracoderma*, and *Anthracobia melaloma* (*9-10,22*). These represent α-1,4-glucan lyases that have been characterized. Even though putative glucan lyase genes are known in cyanobacteria and other fungal species from sequenced genomes (*29*), further research is needed to extend the occurrence of glucan lyase from eukaryotes to prokaryotes with certainty. At the amino acid level, the algal lyases have over 75% identity, and this also applies to the fungal lyases we have cloned (*24-25*). In comparison, the identity between two groups of lyases, *i.e.*, the algal and fungal lyases is only ~20% (*24-25*).

Heterologous expression of lyases has been achieved in selected production organisms of *Aspergillus niger, Pichia pastoris,* and *Hansenula polymorpha* (*22-23*). Among all the eleven glucan lyases studied (*24-25*), we concluded that the red algal glucan lyase (*i.e.,* GLq1-lyase) is of the greatest interest, because its performance is more suitable for the industrial production of AF, and the highest expression yield was achieved in *Hansenula polymorpha* (Yu and Madrid, unpublished results).

Mechanistically, the glucan lyase cleaves the *carbon-oxygen* bond of neighboring D-glucose units in starch, and introduces a double bond between C1 and C2, yielding the 1,2-enol AF that is in equilibration with the ketoform (Figure 1) (*8,24-28*). In aqueous solution, AF exists predominantly in the hydrated form (Figure 1) (*8,44*). In contrast, all other polysaccharide lyases currently known, cleave the *oxygen-carbon* bond and introduce a double bond between C4 and C5, resulting in an oligosaccharide with 4-deoxy 4-enuronosyl group on the non-reducing end (*24*). This further indicates that glucan lyase is substantially different from other lyases in their catalytic mechanism. Our studies indicate that the mechanism of the glucan lyase is similar to α-glucosidase belonging to family 31, by having the same catalytic nucleophile of aspartic acid residue and conserved sequences at the active site (*25-29*). This explains their shared substrate and inhibitor specificity (*8-9*). This hypothesis is further supported by the fact that α-glucosidase catalyzes the protonation of D-glucal, an analogue of the 1,2-enol form of AF, which is virtually the reverse reaction of glucan lyase. These discoveries have provided links between hydrolytic enzymes (EC 3.-.-.-) such as α-glucosidases with the lyase enzymes (EC 4.-.-.-) (*25*), that were, generally, previously regarded as two completely, unrelated groups of starch degrading enzymes.

Substrate Specificity of α-1,4-Glucan Lyase

Except for the difference in products formed, glucan lyase is, to some extent, similar to the exo-acting starch hydrolase β-amylase with respect to substrate specificity. That is, the glucan lyases from both red algae and fungi are highly specific for α-1,4-glucosidic bonds. They show very little activity toward other types of glucosidic bonds, such as α-1,6-glucosidic bonds that also occur in amylopectin and glycogen, and are cleaved only after a long reaction time. The surroundings of α-1,6-glucosidic bonds in a substrate molecule is also important for its cleavage, *e.g.*, the α-1,6-glucosidic bond in the disaccharide isomaltose is much more resistant to cleavage by the red algal GLq1-lyase, than the α-1,6-glucosidic bond in the trisaccharide panose (*8*). The lyase is an exo-acting enzyme that releases AF from the non-reducing ends of starch and glycogen chains (Figure 1). Differences in the substrate specificity between the algal and fungal lyases are that fungal lyases are less active toward maltose and maltodextrins than starch and glycogen, while algal lyases show almost equal activity towards maltose and starch. Algal lyases produce higher yields of AF and render them more suitable for industrial production from starch (*8-9*).

The Stability of Glucan Lyase and its Temperature and pH Optima

The algal glucan lyase from *Gracilariopsis lemaneiformis* (*i.e.*, GLq1-lyase) is stable at 5 °C for at least five years, and even for six months at 22 °C in dilute solution at pH 5.3. GLq1-lyase is also highly resistant to proteolysis, as tested by various proteases (*29*). The pH optimum of the algal lyase is between pH 4.0 and pH 7.0 (*6*), whereas that of the fungal lyases is much narrower at pH 6.5 (*9*). The temperature optima for both the algal and fungal lyases are between 37 and 50 °C, depending on the type of substrate, at a reaction time of 60 min and a substrate concentration of 2% (w/v) (*6,9*). The optimal temperature of the lyase increases with increased substrate concentrations, which is similar for many other enzymes. The algal lyase is also more stable and active in slightly hydrophobic media, such as when low concentrations of ethanol, propanol, and butanol are present (*24*).

Yields of 1,5-Anhydro-D-Fructose and the Effect of Debranching Enzymes

The action of the glucan lyase on amylopectin or glycogen, leads to the formation of a limit dextrin, which we named as neo-limit dextrin to distinguish it from the limit dextrins formed by β-amylase and phosphorylase (neither of these two enzymes are able to break the α-1,6 glucosidic linkages). The yield of AF from potato amylopectin or oyster glycogen is 57-58%, and is increased to over 80% when a debranching enzyme such as pullulanase is present since, as previously mentioned, α-1,4-glucan lyase is highly specific toward α-1,4-glucosidic linkages while amylopectin in starch is linked by both α-1,4- and α-1,6-linkages. A complete degradation of starch requires the co-application of both lyase and debranching enzymes. Both pullulanase (which is more specific for short-branched chains) and isoamylase (which has a preference for longer branched glucans) are efficient in increasing the AF yield. When applying isoamylase, care must be taken, because it will debranch amylopectin and produce insoluble, short chain amylose, which is not an optimal substrate for the lyase.

Production of 1,5-Anhydro-D-Fructose

Historical Review

Chemical synthesis of AF was first reported by Prof. Lichtenthaler and co-workers at Darmstadt Technical University, Germany, in 1980 (*45*). Enzymatic preparation of AF was reported by Prof. Akanuma and coworkers in 1986 by the oxidation of 1,5-anhydro-D-glucitol (AG) using an extract containing pyranose 2-oxidase (*40*). Multiple steps are involved in the chemical synthesis of AF. In comparison, in the enzymatic synthesis of AF there is a quantitative conversion when pure enzyme is utilized, but AG is relatively expensive compared to the cost of chemical synthesis. In 1988, Prof. Baute and coworkers at Bordeaux, France, prepared AF from starch by using an extract from the edible morels (*Morchella vulgaris* and *Morchella costata*) containing α-1,4-glucan lyase (*17*). For the first time, a yield of 400-500 mg of AF was obtained, but the purity was only 90% as a non-purified enzyme preparation was used which contained, among others, enzymes releasing glucose from starch. Due to the limited availability of the lyase enzyme, little research was undertaken on the chemistry and application of AF. Kilogram scale quantities of AF with a purity of 99%, have been produced in our laboratory, using the algal GLq1-lyase heterologously expressed in the fungi *A. niger* and *H. polymorpha*. This has made it possible to study AF as a bulking agent for versatile applications in both food and non-food areas. Even though the chemical methods for making AF from glucose might be improved, it will be very difficult to compete with this efficient enzymatic process, using the red algal glucan lyase.

Production of AF using Immobilized Glucan Lyase is Not Cost-Efficient

Immobilized glucan lyase is especially suitable for the use of dextrins as the starting substrate for AF production. Immobilization of glucan lyase can be achieved using traditional immobilization techniques. We immobilized the red algal GLq1-lyase by using succinimide-activated Sepharose (Affigel 15 gel, Bio-Rad Laboratories). The recovery of lyase activity after immobilization on Affigel 15 gel varied between 40-50%. The operational stability of the column packed with the immobilized lyase was stable for at least 16 days when operated at 24 °C and pH 5.5. However, AF production using immobilized glucan lyase is not, at this present time, cost-efficient because of the lower AF yield and extra costs for preparation of the immobilized enzyme.

Production of AF using Granular or Liquid Forms of Glucan Lyase

When maltodextrin with a DE of 3-10 was used as the substrate at 20-25% (w/v) solid matter, the AF yield was 70-75%. When maize amylopectin was the substrate at 20-25% solid matter, the AF yield was 50-70%. In a typical production process, ~50-70% AF is produced after 2 to 7 days reaction at pH 4.0, in a reactor with 20-25% solid mater of either dextrin (DE 3-10) or maize amylopectin at 24 to 35 °C. Dosing with pullulanase or isoamylase further increases the AF yield. The syrup obtained can be either used directly or spray-dried. The AF produced can be purified by ultra-filtration through a membrane with a 10 KDa cut-off, resulting in an AF purity of 99% if amylopectin is used. Figure 2 below illustrates the saccharification of maize amylopectin paste by using the algal GLq1-lyase expressed in the yeast *H. polymorpha.*

Toxicology and Metabolism of 1,5-Anhydro-D-Fructose

Although mushrooms such as morels of *Morchella* species, and red algae such *Gracilariales* species, can be consumed by humans, and the occurrence of AF from these organisms has been documented (*16-17,2,42*), a better understanding of AF with respect to its toxicology and metabolism was necessary. With our collaborators, we undertook a systematic study of this subject. The results indicated that AF does not cause mutagenesis in either

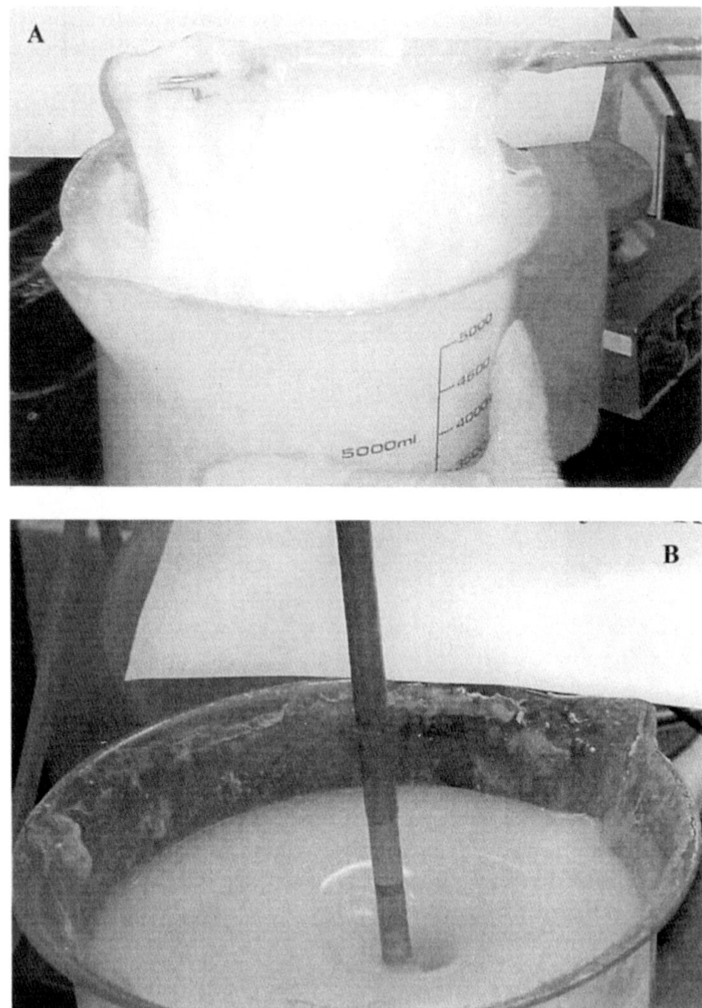

Figure 2. Sacchrification of maize amylopectin by a-1,4-glucan layse to 1,5-anhydrofructose (AF). (A). Before dosing of the glucan layse. (B). 3 days after the dosing of the algal GLq1-glucan lyase.
(See page 1 of color inserts.)

prokaryotic or eukaryotic organisms (*38*). AF at a single dose of 5 g/kg body weight did not produce any adverse effects in rats in a single dose of acute oral toxicity test (*38*). AF given for 90 days at a dose level of 1 g/kg body weight per day also did not cause any observable and measurable symptoms (*39*). Using radioactive labeled AF, it was shown that AF was efficiently absorbed after oral ingestion. Following the intake, AF was reduced in the liver to 1,5-anhydro-D-glucitol (AG) by AF reductase and subsequently secreted in the urine (*38*). These studies were recently documented in the journals of *Food and Chemical Toxicity* and *Drug and Chemical Toxicity* (*38-39*). Ascopyrone P (APP) may have a LD_{50} of 250 mg/kg in rats, and recently it has been found in toasted bread (APP is produced from starch by toasting and early studies have shown that APP can be produced by the pyrolysis of glucose polymers of starch and cellulose). Toxicity studies of microthecin are also still needed.

Conversion of 1,5-Anhydro-D-Fructose to Ascopyrone P and Microthecin

AF is not the final product of starch and glycogen catabolism as shown in Figure 1. It is metabolized to various other metabolite derivatives and the pattern of AF metabolism varies with taxonomic groups. Thus, in red algae such as *G. lemaneiformis* (*2,42*), and in certain fungi such as *M. costata* and *M. vulgaris* (11,16), AF formed by glucan lyase is further converted to microthecin catalyzed by AF dehydratase or aldos-2-ulose dehydratase (Figure 1). In certain other fungi, such as *A. melaloma*, the AF formed is converted by a 2-step reaction to APP via the intermediate ascopyrone M (APM) catalyzed by an AF dehydratase and an enolone tautomerase (Figure 1) (*10,18*). The two AF dehydratases differ from each other. The AF dehydratase from *A. melaloma* is specific for AF, but not for glucosone, an analogue of AF (*10*). The AF dehydratase from the fungus *Phanerochaete chrysosporium* (also named aldos-2-ulose dehydratase, pyranosone dehydratase) is active toward both AF and glucosone and has also an isomerizing activity that isomerizes ascopyrone M to microthecin (Figure 1) (*11*).

Possible Application Areas of 1,5-Anhydro-Fructose and its Derivatives Microthecin and Ascopyrone P

AF and its syrup may have both food and non-food applications, although many of these application areas remain to be explored. Some application areas are highlighted below. For an updated and detailed list, the reader is referred to the website: http://ep.espacenet.com/ (key word is "anhydrofructose"). There are approximately 50 patent applications at this present time.

- AF may be used as an anti-blood clotting reagent (*46*).
- AF and its derivative APP could be antitumor reagents (*47*).
- AF can be used as a humectant.
- AF may be used as an amino group modifier for proteins to increase its hydrophilicity (*49*).
- In contrast to other sugars, AF is basically calorie-free (*38*), with sweetness of less than 20% of sucrose. It might also be used as a filler for certain food products.
- AF displays an antibrowning effect in green tea that is *not* the result of its conversion to APP (*37*) (APP also has an antibrowning effect for green tea but at a lower dosage than AF [*37]*).
- AF reduced the turbidity development in black currant wine during storage (*37*).
- AF has an antimicrobial effect and, therefore, a synergic effect with other antimicrobials. However, this effect of AF may be due to its conversion to APP during the period of food processing and storage (*34-35,48*).
- AF can be a starting material for the manufacturing of certain fine chemicals (*1-5*).
- The AF derivative, microthecin, may be used as an agricultural chemical for plant protection to control fungal pathogens, and for seed coatings to control root rotting diseases caused by *Oomycetes* (*36*).
- The AF derivative, APP may be used as an antioxidant for industrial applications (*37,48*).
- AF could be used to produce novel AF-based oligosaccharides (*3,50*).
- AF may be used to synthesize glycosidase inhibitors including 1-deoxymannojirimycin (*51*).
- AF may be used in the synthesis of novel polymers (*4*).

References

1. Yu, S. *Zuckerindustrie* **2004**, *129*, 26-30.
2. Andersen, S. M.; Lundt, I.; Marcussen, J.; Yu, S. *Carbohydr. Res.* **2002,** *337*, 873-890.
3. Richard, G.; Yu, S.; Monsan, P.; Remaud-Simeon, M.; Morel, S. *Carbohydr. Res.* **2005**, *340,* 395-401.

4. Deppe, O.; Glümer, A; Yu, S.; Buchholz, K. *Carbohydr. Res.* **2004**, *339*, 2077-2082.
5. Andersen, S. M.; Lundt, I.; Marcussen, J. *J. Carbohydr. Chem.* **2000**, *19*, 717-725.
6. Yu, S.; Kenne, L.; Pedersen, M. *Biochim. Biophys. Acta* **1993**, *1156*, 313-320.
7. Yu, S.; Pedersén, M. *Planta* **1993**, *191*, 137-142.
8. Yu, S.; Ahmad, T.; Pedersén, M.; Kenne, L. *Biochim. Biophys. Acta* **1995**, *1244*, 1-9.
9. Yu, S.; Christensen, T. M.; Kragh, K. M.; Bojsen, K.; Marcussen, J. *Biochim. Biophys. Acta* **1997**, *1339*, 311-320.
10. Yu, S.; Refdahl, C.; Lundt, I. *Biochim. Biophys. Acta* **2004**, *1672*, 120-129.
11. Yu, S. *Biochim. Biophys. Acta* **2005**, *1723*, 63-73.
12. Kühn, A.; Yu, S.; Giffhorn, F. *Appl. Environ. Microbiol.* **2006**, *72*, 148-1257.
13. Yu, S.; Blennow, A.; Bojko, M.; Madsen, F.; Olsen, C.-E.; Engelsen, S. B. *Starch/Stärke* **2002**, *54*, 66-74.
14. Shiga, Y.; Kametani, S.; Mizuno, H.; Akanuma, H. *J. Biochem.* (Tokyo) **1996**, *119*, 173-179.
15. Shiga, Y.; Kametani, S.; Kadokura, T.; Akanuma, H. *J. Biochem.* (Tokyo) **1999**, *125*, 166-172.
16. Deffieux, G.; Baute, R.; Baute, M.-A.; Atfani, M.; Carpy, A. 1,5-*Phytochemistry* **1986**, *26,* 1391-1393.
17. Baute, M.-A.; Baute, R.; Deffieux, G. *Phytochemistry* **1988**, *27*, 3401-3403.
18. Baute, M.-A.; Deffieux, G.; Vercauteren, J.; Baute, R.; Badoc, A. *Phytochemistry* **1993**, *33*, 41-45.
19. Kametani, S.; Shiga, Y.; Akanuma, H. *Eur. J. Biochem.* **1996**, *242,* 832-838.
20. Suzuki, M.; Kametani, S.; Uchida, K.; Akanuma, H. *Eur. J. Biochem.* **1996**, *240*, 23-29.
21. Sakuma, M.; Kametani, S.; Akanuma, H. *Biochem.* (Tokyo) **1998**, *123*, 189-193.
22. Bojsen, K.; Yu, S.; Marcussen, J. *Plant Mol. Biol.* **1999**, *40*, 445-454.
23. Bojsen, B.; Yu, S.; Kragh, K. M.; Marcussen, J. *Biochim. Biophys. Acta* **1999**, *1430*, 396-402.
24. Yu, S.; Marcussen, J. In *Recent Advances in Carbohydrate Bioengineering;* Gilbert, H. J.; Davies, G. J.; Henrissat, B.; Svensson, B., Eds.; Royal Society of Chemistry (RS.C): London, *1999*; p 243-250.
25. Yu, S.; Bojsen, B.; Svensson, B.; Marcussen, J. *Biochim. Biophys. Acta* **1999**, *1433*, 1-15.
26. Lee, S. S.; Yu, S.; Withers, S. G. *J. Am. Chem. Soc.* **2002**, *124*, 4948-4949.
27. Lee, S. S.; Yu, S.; Withers, S. G. *Biochemistry* **2003**, *42*, 1381-1390.

28. Lee, S. S.; Yu, S.; Withers, S. G. *Biologia* **2005**, *60(16)*, 137-148.
29. Ernst, H. A.; Leggio, L. L.; Yu, S.; Finnie, C.; Svensson, B.; Larsen, S. *Biologia* **2005**, *60(16)*, 149-159.
30. Andersen, S. M.; Lundt, I.; Marcussen, J.; S¢tofte, I.; Yu, S. *J. Carbohydr. Chem.* **1998**, *17*, 1027-1035.
31. Yu, S.; Olsen, C. E.; Marcussen, J. *Carbohydr. Res.* **1998** , *305*, 73-82.
32. Andersen, S. M.; Jensen, H. M.; Yu, S. *J. Carbohydr. Chem.* **2002**, *21*, 569-578.
33. Ahrén, B.; Holst, J.J.; Yu, S. *Eur. J. Pharmacol.* **2000**, *397,* 219-225.
34. Thomas, L. V.; Yu, S.; Ingram, R. E.; Refdahl, C.; Elsser, E.; Delves-Broughton, J. *J. Appl. Microbiol.* **2002**, *93*, 697-705.
35. Thomas, L. V.; Ingram, R. E.; Yu, S., Delves-Broughton, J. *Intl. J. Food Microbiol.* **2004**, *93*, 319-323.
36. Morgan, A. J.; Turner, M.; Yu, S.; Weiergang, I., Pedersen, H. C. *Intl. Patent Appl.* WO2004083226 **2004**, p 1-53
37. Yuan, Y. B.; Mo, S. X.; Cao, R.; Westh, B.; Yu, S. *J. Agric. Food Chem.* **2005**, *53*, 9491–9497.
38. Yu, S.; Mei, J.; Ahrén, B. *Food Chem. Toxicol.* **2004**, *42*, 1677-1686.
39. Mei, J.; Yu, S.; Ahrén, B. *Drug Chem. Toxicol.* **2005**, *28,* 263-272.
40. Nakamura, T.; Naito, A.; Takahashi, Y.; Akanuma, H. *J. Biochem.* (Tokyo) **1986**, *99*, 607-613.
41. Hirano, K.; Ziak, M.; Kamoshita, K.; Sukenaga, Y.; Kametani, S.; Shiga, Y.; Roth, J.; Akanuma, H. N. *Glycobiology* **2000**, *10*, 1283-1289.
42. Broberg, A.; Kenne, L.; Pedersén, M. *Anal. Biochem.* **1999**, *268*, 35-42.
43. Yoshinaga, K.; Fujisue, M.; Abe, J.-I.; Hanashiro, I.; Takeda, Y.; Muroya, K.; Hizukuri, S. *Biochim. Biphys. Acta* **1999**, *1472*, 447-454.
44. Taguchi, T.; Haruna, M.; Okuda, J. *Biotechnol. Appl. Biochem.* **1993**, *18*, 275-283.
45. Lichtenthaler, F. W.; El Ashry, E. S. H.; Göckel, V. H. *Tetrahedron Lett.* **1980**, *21*, 1429-1432.
46. Maruyama, I.; Hizukuri, S.; Yamaji, K. *Intl. Patent Appl.* WO2004045628, **2004**, p 1-23.
47. Maruyama, Y.; Abeyama, K.; Yoshimoto, Y. Antitumoral agent. *Intl. Patent Appl.* WO2005040147, **2005**, p 1-27
48. Yoshinaga, k.; Wakamatsu, C.; Saeki, Y.; Abe, J.-I.; Hizukuri, S. *J. Appl. Glycosci.* **2005**, *52*, 287-291.
49. Yoshinaga, K.; Fujisue, M.; Abe, J. I.; Takeda, Y.; Hizukuri, S. *J. Appl. Glycosci.* **2002**, *49*, 129-135.
50. Yoshinaga, K.; Abe, J.-I.; Tanimoto,Y.; Koizumi, K.; Hizukuri, S. *Carbohydr. Res.* **338**, *2003*, 2221-2225.
51. Maier, P.; Andersen, S. M.; Lundt, I. *Synthesis* **5**, *2006*, 827-830.

Chapter 5

Stabilization and Activation of Nine Starch Degrading Enzymes and Significant Differences in the Activities of α-Amylases Assayed on Eight Different Starches

John F. Robyt

Laboratory of Carbohydrate Chemistry and Enzymology, Department of Biochemistry, Biophysics, and Molecular Biology, Iowa State University, Ames, IA 50011

The stabilization and activation of nine starch-degrading enzymes have been studied using nonionic detergents, Triton X-100, seven polyethylene glycols of different average molecular weights, 400–8K Da, and two polyvinyl alcohols of 10K and 50K Da. Most of the additives gave activation, but there was a specific additive that gave the maximum degree of activation. Assay of *Bacillus amyloliquefaciens* and porcine pancreatic α-amylases on eight different starches were found to be significantly different. The activities were found to be dependent on the method used to solubilize the starches. The activities were also found to be dramatically increased to 1.2–9.1 times by the addition of polyethylene glycols 1.0K and 1.5K to the enzymes.

Introduction

Enzymes often undergo a time-dependent inactivation, following dilution to low concentrations that are necessary for assay (*1-2*) for reasons that include surface denaturation (*1-3*), presence of inactivators (*4*), and dissociation of active oligomers into inactive subunits (*5*). One method commonly used to prevent these inactivations is the addition of relatively large amounts (1–2 mg/mL) of an enzymatically inert protein, such as bovine serum albumin (*1*), but this is not always effective or desirable (*6*). Another method is to add ligands, such as substrates, products, or effectors of the enzymes to the enzyme solution (*2,7*), which also are frequently undesirable, especially in analytical, preparative, or industrial applications. We, thus, sought the addition of other agents that can be added in low concentrations and have little or no effect on the enzyme catalyzed reactions, except to stabilize the enzyme. We found that nonionic detergents, such as 0.02%–0.04% (w/v) Triton X–100, seven polyethylene glycols (PEGs) of average MWs of 400–8K and two polyvinyl alcohols (PVAs) of 10K and 50K were stabilizers for nine starch-degrading enzymes that form diverse products. In addition to stabilization, we found that they also gave significant degrees of activation from 20 to 77%. We also found that the activities of two α-amylases were significantly different when assayed on eight different starches and that the activities were dependent on how the starches were solubilized. Further, the addition of PEG 1.0K and 1.5K to *Bacillus amyloliquefaciens* α-amylase (BAA) and porcine pancreatic α-amylase (PPA), respectively, gave dramatic increases (1.25–9.1 times) to their activities on the starches. We report here on these finding and give interpretations for the stabilization and activation of the enzymes and the differences in the activities of the two α-amylases on eight different starches in the absence and presence of the PEGs.

Stabilization and Activation of Nine Starch-degrading Enzymes

It was observed that a dilute solution of PPA under optimum conditions of pH 6.5, 20 °C, and 1 mM $CaCl_2$ lost 98% of its activity over a 2 h period (Figure 1). Previous experience with the addition of a nonionic detergent, Triton X–100, to a solution of dextransucrase gave stabilization (*6*). On the addition of 0.02% (w/v) Triton X–100 to a freshly diluted solution of PPA, the enzyme was stabilized but, much to our surprise, it also was significantly activated to the extent of 41% (*8*), which is illustrated in Figure 1.

Examination of the structure of Triton X–100 showed that it had a polyethylene glycol side chain of 10–11 monomer units attached to a benzene

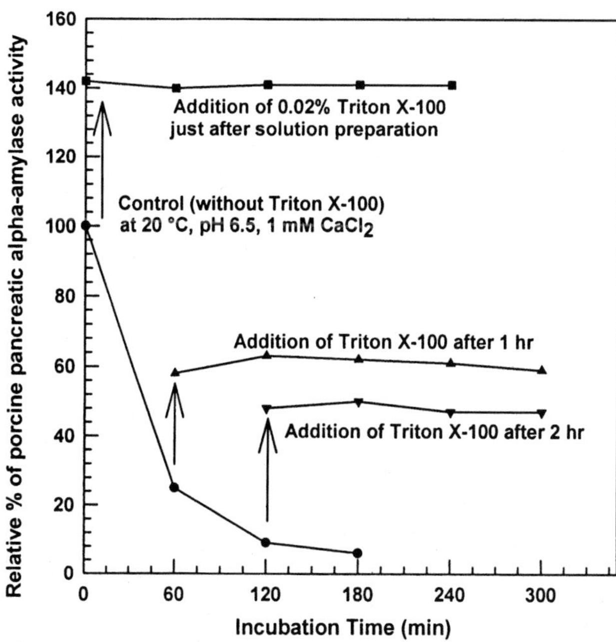

Figure 1. Effect of adding 0.02% (w/v) Trition X-100 on the activity of porcine pancreatic α-amylase (PPA); loss of dilute PPA activity on standing (λ); stabilization and activation on addition of 0.02% (w/v) Triton X–100 (ν); reactivation and stabilization after loss of activity for 1h (σ); and after standing for 2 h (τ). (Adapted with permission from reference 8. Copyright 2005 Elsevier Press).

ring. We decided to examine the stabilization and activation of PPA by seven polyethylene glycols with the average molecular weights of 400–8,000 Da and by two polyvinyl alcohols with average molecular weights of 10K and 50K Da. We found that nearly all of them gave stabilization and activation of PPA. There was one, however, 0.02% (w/v) PEG 1.5K that gave a mazimum degree of activation of 54% for PPA (Table I).

Because there are technological applications and theoretical interests in the action of various kinds of starch-degrading enzymes that produce different types of products, nine different starch-degrading enzymes were examined: five different α-amylases, from porcine pancreas (PPA), human saliva (HSA), *Aspergillus oryzae* (AOA), *Bacillus licheniformis* (BLA), *Bacillus amyloliquefaciens* (BAA); and four other enzymes: barley β-amylase (β-A), *A. niger* glucoamylase (GA), *Pseudomonas amylodermosa* isoamylase (IA), and *Bac. macerans* cyclomaltodextrin glucanyltransferase (CGT). The activities of

Table I. Activation of Nine Starch-degrading Enzymes by 0.02% (w/v) Triton X–100, Seven PEGs, and Two PVAs[a]

	Relative percent activity								
Additives	*PPA* %	*HSA* %	*AOA* %	*BLA* %	*BAA* %	*β-A* %	*GA* %	*IA* %	*CGT* %
control	100	100	100	100	100	100	100	100	100
TritonX-100	141	**145**	132	135	**134**	155	118	130	108
PVA 10K	136	140	127	131	128	136	120	123	103
PVA 50K	128	142	133	135	121	121	**130**	121	102
PEG 400	100	86	134	119	105	107	103	102	96
PEG 600	127	133	136	124	116	119	107	100	101
PEG 1.0K	145	143	138	136	124	160	108	119	102
PEG 1.5K	**154**	139	**142**	143	123	**177**	121	132	**120**
PEG 2.0K	144	134	137	**144**	119	158	117	**137**	96
PEG 4.6K	141	129	134	**144**	119	174	110	136	91
PEG 8.0K	138	140	134	140	119	168	111	134	90

[a] The maximum activity for each enzyme is given in bold type.

The data are from reference (*8*).

PPA, HSA, AOA, BLA, and BAA were determined on 'soluble amylose'; the activities of β-A, GA, IA were determined on waxy maize starch. Samples were removed every 5 min for 30 min and the reducing value was measured by the micro copper bicinchoninate method, using maltose as the standard (*10*). The unit of activity (U) was 1.0 μmole of glycosidic bond hydrolyzed per min obtained from the slope of the linear line of μg of maltose *vs* time (min) divided by 342, the molecular weight of maltose. The activity of GA was obtained by measuring the amount of D–glucose produced per min, as determined by the micro glucose oxidase method (*10*). The activity of CGTase was obtained from the reaction of cyclomaltohexaose plus methyl α-D–glucopyranoside, followed by reaction with GA, and the amount of D–glucose produced was measured by the micro glucose oxidase method (*10-11*).

Figure 1 shows the loss of PPA activity over a 2 h period, when standing in a dilute solution, equivalent to a concentration used for assay. The addition of 0.02% (w/v) Triton X–100, just as the enzyme was diluted, gave stabilization over 4 h, but also gave 41% activation. After 1 h of standing, Triton X–100 was added and the enzyme regained ~26% activity, with stabilization; addition of Triton X–100 after standing 2 h gave ~30% activity, with stabilization. In neither case, however, did the enzyme regain anything close to its original activity.

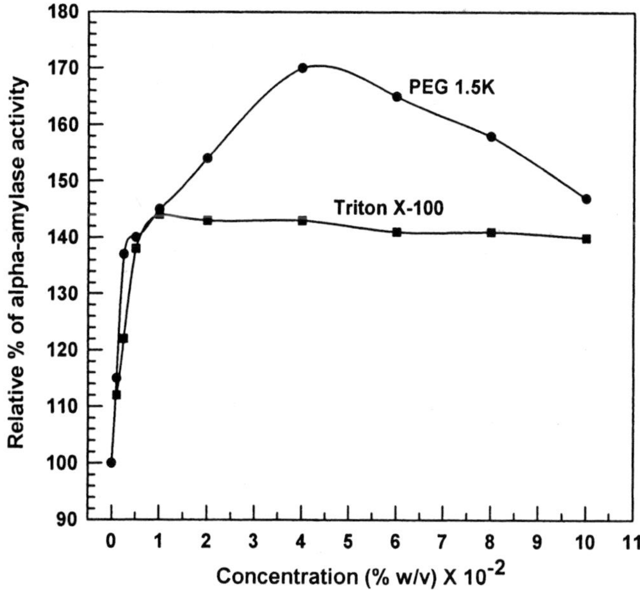

Figure 2. Effects of the concentration of Triton X–100 (ν) and PEG 1.5K (λ) on the activity of porcine pancreatic α-amylase. (Adpated with permission from reference 8. Copyright 2005 Elsevier Press).

The effects of the concentration of Triton X–100 and PEG 1.5K for PPA are shown in Figure 2. When PEG 1.5K was added, the maximum amount of activity (70%) was obtained at 0.04% (w/v). The effects of adding activators (Triton X–100 and PEG 1.5K) to (a) only the substrate solution, (b) only the dilute enzyme solution, (c) both the enzyme and the substrate solutions were studied is shown in Figure 3. These results show that an equal amount of activation was obtained for cases (b) and (c), but in case (a) where the activator was only added to the substrate and not to the enzyme, no activation was obtained. These experiments show that the effect of the activator was on the enzyme and not on the substrate.

The activities of the nine starch-degrading enzymes when 0.02% (w/v) Triton X–100, seven PEGs, and two PVAs are given in Table I. Table I shows that different additives give different degrees of activation and in a few cases inhibition. Triton X–100 gave the maximum degree of activation for two enzymes (HSA and BAA); PEG 1.5K gave the maximum degree of activation for four enzymes (PPA, AOA, β-A, and CGT); PEG 2.0K gave maximum activation for two enzymes (BLA and IA); and PVA 50K gave maximum activation for one enzyme (GA). Doubling the concentration of the activators to

0.04% (w/v), however, significantly increased the activity for six of the enzymes and changed the activator from Triton X–100 to PEG 1.0K for BAA that gave the maximum degree of activation, but it also decreased the activities for three of the enzymes (β-A, HSA, and CGT) and 0.02% gave the maximum degree of activation for these enzymes (Table II).

The maximum degree of activation was increased from 54 to 70% for PPA on addition of 0.04% PEG 1.5K; the activity of IA was increased from 37 to 58% on addition of 0.04% PEG 2.0K; the activity of AOA was increased from 42 to 53% on addition of 0.04% PEG 1.5K; the activity of GA was increased from 30 to 48% on the addition of 0.04% PVA 50K; the activity of BLA was increased from 44 to 46% on the addition of 0.04% PEG 4.6K; and the activity of BAA was increased from 34% on the addition of 0.02% Triton X-100 to 42% on the addition of 0.04% PEG 1.0K.

It is postulated, that the various enzymes in dilute solution exist in equilibrium in several different tertiary forms, each having a different amount of enzyme activity. The mechanism of activation and stabilization of the nine starch-degrading enzymes by the various maximum activating agents is their binding to the enzyme-protein forms to give a single optimally folded tertiary structural form that has the maximum enzyme activity. The continued binding of the additives by a strong interaction with the enzyme-protein produces stabilization as well as activation. In the experiment illustrated in Figure 3, in which the additive was only added to the substrate solution, there was no activation, but in the experiment where the additive was added to enzyme and to both the enzyme and the substrate, there was an equal percent activation of PPA. This shows that the effect of the additive is exclusively on the enzyme-protein and not on the substrate.

Significant Differences in the Activities of α-Amylases in the Absence and Presence of Polyethylene Glycols Assayed on Eight Different Starches

It is usually believed that the assay of a single α-amylase on different starches would be the same. Recently, in assaying *Bacillus amyloliquefaciens* α-amylase (BAA) on potato starch and on waxy maize starch, it was surprisingly found that the two starches gave significantly different assay values. We then assayed porcine pancreatic α-amylase (PPA) on the two starches and also found significantly different assay values.

Because of these results, we then assayed the two amylases on eight different starches from potato, rice, wheat, maize, and tapioca, the major food starches; from amylomaize-7, a high amylose starch with 70% amylose, and waxy maize, a high amylopectin starch with 100% amylopectin; and shoti

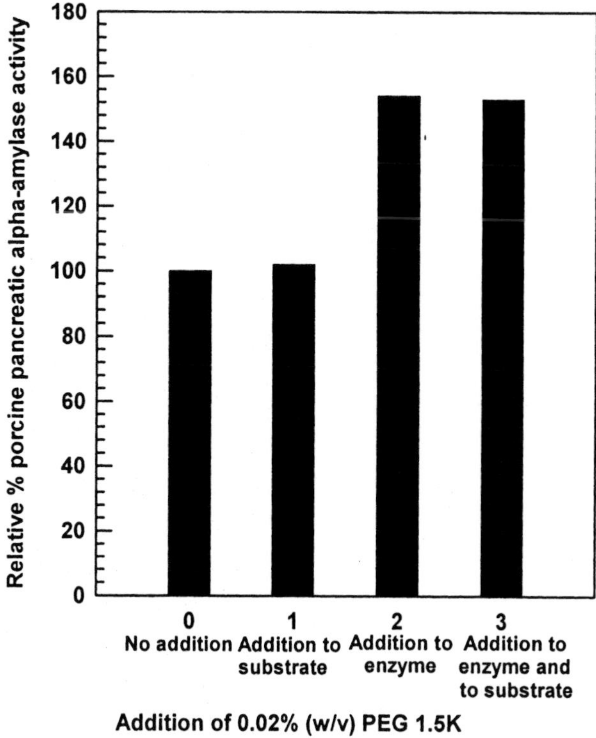

Figure 3. Effects of adding 0.02% (w/v) PEG 1.5K (1) to only the substrate, (2) to only PPA, and (3) to both PPA and the substrate.

Table II. Additives and Their Concentrations Giving the Maximum Degree of Activation for the Nine Starch-Degrading Enzymes

Enzyme	*Maxium % activation*	*Additive % (w/v)*	*Additive*
β-A	77	0.02	PEG 1.5K Da
PPA	70	0.04	PEG 1.5K Da
IA	58	0.04	PEG 2.0K Da
AOA	53	0.04	PEG 1.5K Da
GA	48	0.04	PVA 50K Da
BLA	46	0.04	PEG 4.6K Da
HSA	45	0.02	Triton X–100
BAA	42	0.04	PEG 1.0K Da
CGT	20	0.02	PEG 1.5K Da

Data from reference (*8*).

starch, which is a somewhat exotic tuber starch with a flat plate granule morphology and is used as a therapeutic agent in Asia for various stomach and intestinal problems. Furthermore, potato, tapioca, and shoti starches are also relatively easily gelatinized, while maize, rice, amylomaize-7, and wheat starches are relatively difficult to gelatinize.

The starches were solubilized by suspending 110 mg of each starch in 7 mL of water and autoclaving them at 121 °C for 30 min; solubilization did not produce depolymerization. The autoclaved starches were centrifuged, buffer was added, and the solutions were diluted with water to 10.0 mL. The enzyme reactions were initiated by the addition of 100 µL of freshly diluted enzyme to 1.90 mL of starch solution with 40 mM buffer (pH 6.5), containing 1 mM $CaCl_2$ all of which was preincubated at 37 °C; 100 µL aliquots were taken every 5 min for 30 min and the reducing value was measured by the micro copper bicinchoninate method, using maltose at the standard (*10*). The unit of activity was determined as previously described in the section on the stabilization and activation of the nine starch-degrading enzymes. Because PEGs gave significant degrees of activation, 0.04% (w/v) PEG 1.0K was added to BAA and 0.04% (w/v) PEG 1.5K was added to PPA and they were assayed on the eight starches solubilized by the two methods: (a) the autoclaving method and (b) stirring the starch in 1 M NaOH for 15 h at 20 °C. In these experiments, it was found that there was a dramatic significant increase in the activities of the enzymes assayed on the starches solubilized by the two methods (see Tables III and IV). The addition of the PEGs to BAA and PPA gave large increases when they were assayed on the eight starches solubilized by the two methods.

In most cases the maximum activities for the two enzymes were obtained from the starches solubilized by 1 M NaOH. Table V shows a comparison of the order of the activities of the two enzymes on the solubilization by 1 M NaOH. The order of activity for the first three starches were the same for the two enzymes, but in reverse order: tapioca, potato, and shoti for BAA and shoti, potato, and tapioca for PPA. Waxy maize starch was fourth for both enzymes. Wheat and rice starches were fifth and sixth for BAA and PPA, but in reverse order. The last two starches, amylomaize-7 and maize, had the lowest activities in the same order for both enzymes.

For both enzymes, BAA and PPA, the assays are significantly different on the eight starches. These differences suggest that they were due to differences in the structures of the starches. It is recognized that starch granules from different sources have varying degrees of molecular and supra-molecular order that is produced by secondary structures (single and double helices of starch chains) and tertiary structure (intra- and inter-molecular associations by hydrogen and hydrophobic bonding) (*12-13*). Starch granules exhibit a wide diversity in their size and morphology, composition of amylose and amylopectin, degrees of crystallinity, gelatinization, temperatures, types of X-ray patterns, and susceptibility to enzyme hydrolysis (*14*). It has been found (*15*) that α–amylases

Table III. Activity of *Bacillus Amyloliquefaciens* α-Amylase in the Absence and Presence of Polyethylene Glycol 1.0K Solubilized by Autoclaving at 121 °C for 30 Min and by 1 M NaOH at 20 °C for 15 h. Reproduced from reference 18 with permission. Copyright 2006, Elsevier Press.

Method of Starch Solubilization	*Autoclaved 121 °C 30 min*		*1 M NaOH 20 °C 15 h*	
Starches	*w/o PEG 1.0K U[a]/mL ± SD[b]*	*w/ PEG 1.0K U[a]/mL ± SD[b]*	*w/o PEG 1.0K U[a]/mL ± SD[b]*	*w/PEG 1.0K U[a]/mL ± SD[b]*
Shoti	317 ± 8	411 ± 8	241 ± 3	802 ± 4
Waxy maize	317 ± 8	408 ± 9	272 ± 5	593 ± 7
Tapioca	293 ± 4	438 ± 6	471 ± 9	1103 ± 8
Rice	288 ± 5	444 ± 7	251 ± 4	455 ± 6
Potato	268 ± 7	405 ± 7	397 ± 3	822 ± 4
Maize	262 ± 5	496 ± 7	291 ± 4	355 ± 4
Wheat	210 ± 3	524 ± 11	292 ± 6	501 ± 4
Amylomaize-7	184 ± 4	411 ± 4	233 ± 4	404 ± 4

[a] U = 1.0 μmole of α-1→4 glycosidic bonds hydrolyzed/min

[b] SD = standard deviation.

Table IV. Activity of Porcine Pancreatic α-Amylase in the Absence and Presence of Polyethylene Glycol 1.5K Solubilized by Autoclaving at 121 °C for 30 Min and by 1 M NaOH for 15 h. Reproduced from Reference 18 with Permission. Copyright 2006, Elsevier Press.

Method of Starch Solubilization	*Autoclaved 121 °C 30 min*		*1 M NaOH 20 °C 15 h*	
Starches	*w/o PEG 1.0K U[a]/mL ± SD[b]*	*w/ PEG 1.0K U[a]/mL ± SD[b]*	*w/o PEG 1.0K U[a]/mL ± SD[b]*	*w/PEG 1.0K U[a]/mL ± SD[b]*
Maize	640 ± 7	905 ± 7	442 ± 6	824 ± 54
Wheat	626 ± 19	1003 ± 21	435 ± 7	864 ± 26
Shoti	613 ± 16	3137 ± 6	1219 ± 12	3462 ± 47
Waxy maize	557 ± 1	2500 ± 26	1141 ± 6	1330 ± 6
Amylomaize-7	523 ± 8	856 ± 30	455 ± 4	858 ± 7
Rice	451 ± 6	798 ± 21	417 ± 23	890 ± 7
Tapioca	281 ± 3	2841 ± 24	1007 ± 9	2450 ± 2
Potato	262 ± 2	2100 ± 1	1274 ± 20	3152 ± 68

[a] One (U) = 1.0 μmole of α-1→4 glycosidic bonds hydrolyzed/min

[b] SD = standard deviation.

Table V. Comparison of the maximum activities of *Bacillus amyloliquefaciens* α-amylase and porcine pancreatic α-amylase in the presence of PEG on the eight starches solubilized by 1 M NaOH. Reproduced from reference 18 with permission. Copyright 2006, Elsevier Press.

Bacillus amyloliquefaciens α-amylase			Porcine pancreatic α-amylase		
Order	*Starches*	*U[a]/mL*	*Order*	*Starches*	*U[a]/mL*
1	Tapioca	1103	1	Shoti	3462
2	Potato	822	2	Potato	3152
3	Shoti	802	3	Tapioca	2450
4	Waxy maize	593	4	Waxy maize	1330
5	Wheat	501	5	Rice	890
6	Rice	455	6	Wheat	864
7	Amylomaize-7	404	7	Amylomaize-7	858
8	Maize	355	8	Maize	824

[a] One (U) = 1.0 μmole of α-1→4 glycosidic bonds hydrolyzed/min

will not hydrolyze the glycosidic linkages of glucose units involved in a double helix, as well as retrograded (associated) starch chains.

It was, thus, postulated that the differences observed in the activities of a single kind of α-amylase, assayed on the solubilized eight starches, was due to the retention in solution of different amounts of secondary and tertiary structures of the starch granules, and that the amounts of these structures varied for the different starches and hence gave different activities for a single kind of α-amylase.

To test this hypothesis, the eight starches were solubilized by another method than autoclaving, *e.g.*, by treatment with 1 M NaOH for 15 h at 20 °C and then assayed. The results showed that the majority of the activities were significantly increased for some of the starches, but also there were some that decreased, indicating that the activities were dependent on the method of solubilization of the starches. The different amounts of activities obtained by the two methods of solubilization are dependent on the mechanism by which the two methods affect the secondary and tertiary structures in the granule and hence give a certain percent retention of different amounts of secondary and tertiary structural features to the starches in solution.

Even though the two enzymes had widely different activities on the various starches, they had comparable levels of activity when assayed on the eight starches, with approximately 2.0- to 3.5 times higher values for PPA over BAA. This reflects differences in the product specificities of the two enzymes, in which BAA gives products that on the average are 2.5-times larger than the products produced by PPA (*16-17*) and hence BAA gives lower reducing values than PPA for a comparable amount of catalysis.

The significantly higher activities that are obtained on the addition of PEGs to the two α-amylases are due to specific affects on the structure of the two enzymes, as previously postulated, while the differences in the activities on the eight starches are postulated to be due to the retention of some of the secondary and tertiary structures of the starches when in solution. For the majority of the starches, the highest activities for the two enzymes were obtained when PEG was added to them and the starches were solubilized by 1 M NaOH. For these two conditions, we see a correlation between the two enzymes for the activities on the starches. The activities of both enzymes were highest on tapioca, potato, shoti, and waxy maize starches in nearly the same order. These starches are more easily gelatinized than the last four starches and hence must have lesser amounts of resistant secondary and tertiary structures when in solution. The last four starches had significantly lower activites in nearly the same order for the two enzymes of wheat, rice, amylomaize-7, and maize starches, all of which are relatively difficult to gelatinize, indicating strong secondary and tertiary structural features and hence higher retention of these structures when in solution. Also the activities of the two α-amylases on the eight starches were significantly higher than they were when assayed with soluble amylose, which has a relatively simple structure and might be expected to have minimal amounts of secondary and tertiary structures when in solution.

References

1. Allison, R. D.; Purich, D. L. *Methods Enzymol.* **1979**, *63*, 3–22.
2. Fromm, H. J. In *Initial Rate Enzyme Kinetics*; Springer-Verlag: Berlin, **1975**; p 43–45.
3. Dixon, M.; Webb, E. C. *Enzymes*; Academic Press: New York, **1979**; p 11–12.
4. Kilbanov, A. M. *Anal. Biochem.* **1979**, *93*, 1–25.
5. Bernfeld, P.; Berkeley, B. J.; Bieber, R. E. *Arch. Biochem. Biophys.* **1965**, *111*, 31–38.
6. Miller, A. W.; Robyt, J. F. *Biochim. Biophys. Acta* **1984**, *78*, 589–596.
7. Wiseman, A. In *Topics in Enzyme and Fermentation Biotechnology*; Wiseman, A.; Ed.; Ellis, Horwood, Chichester, and Halsted Press: New York; **1978**; *Vol. 2*, p 280–303.
8. Yoon, S.-H.; Robyt, J. F. *Enzyme Microbial Technol.* **2005**, *37*, 556–562.
9. Robyt, J. F. In *Starch: Chemistry and Technology*; Whistler, R. J., BeMiller, J. N.; Paschall, E. F.; Eds.; 2nd edition; Academic Press: New York; **1984**; p 87–123.

10. Fox, J. D.; Robyt, J. F. *Anal. Biochem.* **1991**, *195*, 93–96.
11. Lee, S.-B.; Robyt, J. F. *Carbohydr. Res.* **2001**, *336*, 47–53.
12. French, D. In *Starch: Chemistry and Technology*; Whistler, R. J., BeMiller, J. N., Paschall, E. F.;` Eds.; 2nd edition; Academic Press: New York; **1984**; p 184–247.
13. Zobel, H. F. *Starch/Stärke*, **1988**, *40*, 44–50.
14. Zobel, H. F.; Stephen, A. M. In *Food Polysaccharides and Their Applications*; Stephen, A. M., Ed.; Marcel Dekker: New York; **1995**; p 19–66.
15. Jane, J-L.; Robyt, J. F. *Carbohydr. Res.* **1984**, *132*, 105–118.
16. Robyt, J. F.; French, D. *Arch. Biochem. Biophys.* **1963**, *100*, 451–462.
17. Robyt, J. F.; French, D. *J. Biol. Chem.* **1970**, *245*, 3917–3923.
18. Mukerjea, R.; Slocum, G.; Mukerjea, R.; Robyt, J. F. *Carbohydr. Res.* **2006**, *34,* 2049-2054.

Chapter 6

Overcoming Practical Problems of Enzyme Applications in Industrial Processes: Dextranases in the Sugar Industry

Gillian Eggleston[1], Adrian Monge[2], Belisario Montes[3], and David Stewart[3]

[1]Commodity Utilization Research Unit, Southern Regional Research Center, Agricultural Research Service, U.S. Department of Agriculture, New Orleans, LA 70124

[2]Cora Texas factory, P.O. Box 280, White Castle, LA 70788

[3]Alma Plantation LLC, 4612 Alma Road, Lakeland, LA 70752

Dextranases only have a small market and low volume sales compared to many other industrial enzymes. Consequently, research and development efforts to engineer properties of dextranases to specific conditions of industrial processes have not occurred and are not expected soon. This book chapter highlights the difficulties associated with the practical application of dextranases, that are sometimes applied to hydrolyze dextran in sugar manufacture when bacterial deterioration of sugarcane or sugarbeet has occurred. Less than optimum application existed because of confusion about where to add the dextranase in the factory/refinery and which commercial dextranase to use. The wide variation in activity of commercially available dextranases in the U.S., and a standardized titration method to measure activities at the factory are discussed. Optimization by applying "concentrated" dextranase as a working solution to heated juice is described. Promising short-term technologies to further improve industrial dextranase applications are discussed, as well as the long-term outlook.

Introduction

The major contributor to sugarcane and sugarbeet deterioration in the U.S. is *Leuconostoc mesenteroides* infections (*1-2*). Factors affecting infection in the sugarcane plant and extracted juice are ambient temperature and humidity, level of rainfall and mud, length of sugarcane billet, degree of burning, billet damage, delays between burning and cutting and subsequent processing, and mill hygiene. *L. mesenteroides* produce dextrans (α-(1→6)-α-D-glucans), that in moderate and severe cases can interrupt normal processing operations. Dextran causes expensive microbial losses of sucrose and losses of sucrose to molasses. Moreover, the factory is penalized by refineries for dextran in the raw sugar. Dextrans are polydisperse by nature, *i.e.*, they exist as a wide range of molecular weights. The high viscosity associated with the HMW portions (> 1000 KDa) of dextran affects boiling house operations, often reducing rates of evaporation and crystallization. Dextrans isolated from sugarcane products possess a largely linear structure (*3*) comprised of ~95% glucose units linked by (1→6) glycosidic bonds, but also containing ~5% branching through (1→4), (1→3) and some (1→2) linkages (Figure 1).

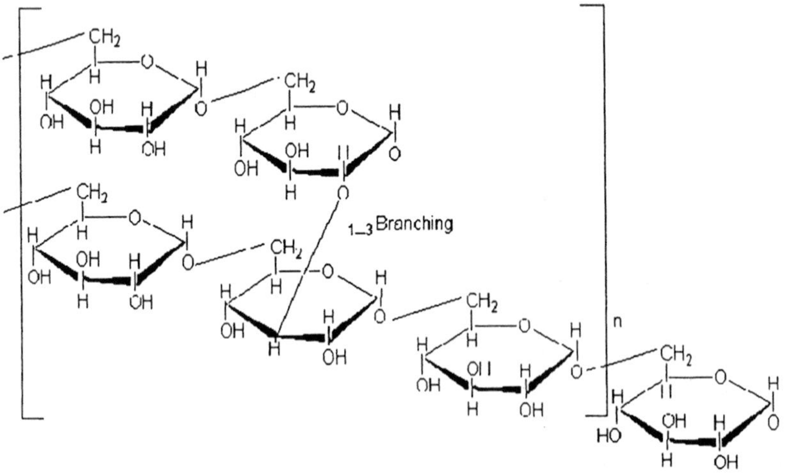

Figure 1. Basic chemical structure of dextran. From Eggleston et al. (4).

Commercial dextranases (1→6-α-glucan hydrolases, EC 3.2.1.11) have been used in sugarcane and sugarbeet (*2*) factories/refineries to break down dextran by hydrolyzing α-(1→6) linkages at random endogenous sites (*3*). The hydrolysis of dextran by dextranase is not an "all or nothing" mechanism. Instead, there is a gradual decrease in the average molecular weight of the various dextran fragments produced from the original HMW dextran, and these fragments in turn are continuously hydrolyzed (Figure 2). Dextranase is mostly added to improve boiling house operations.

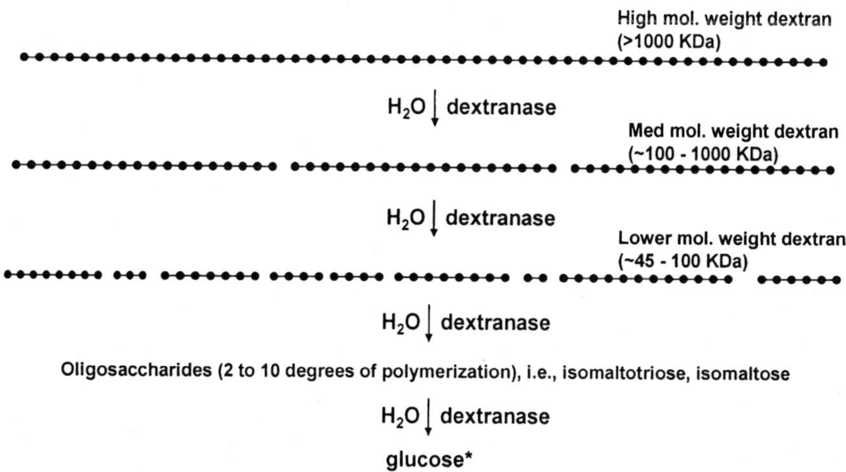

Figure 2. Mode of endohydrolysis action of dextranase on α-(1→6) glycoside linkages in random sites of HMW dextran. Dots and connecting lines represent <u>*chains*</u> *of glucose molecules linked by α-(1→6) bonds in the dextran molecule. Reaction time is not represented. * In dextranase application in the sugar industry, glucose as the end product rarely or never occurs. From Eggleston et al. (4).*

Although the application of dextranases in the sugar industry was pioneered by Australians in the 1970s (*5-7*), their application in U.S. sugar manufacture is still not optimized. This is partly because of confusion about which dextranase to use, and how and where to add the dextranase. The problem has been exacerbated by the lack of comprehensive U.S. dextranase studies, and the few studies that have been undertaken most never stating the dextranase activity (*8-10*). This has led to enormous confusion concerning what is the appropriate dosage to apply for each commercial dextranase. Also, dextranases have a small market and low volume sales compared to other industrial enzymes such as α-amylases in the large detergent and food industries. Thus, they have not been subjected to much research and development by large enzyme companies and their properties have not been tailored to sugar industry conditions. As a consequence, at the request of the Louisiana raw sugar industry, studies were undertaken to optimize the application of dextranases in U.S. factories.

Standard Method to Measure Dextranase Activity at Factories or Refineries

Until recently, the optimization of dextranase applications in sugar factories/refineries was hindered by the lack of a standard or uniform method to measure the activity of commercial dextranases and, unfortunately, there is no regulatory body in the U.S. to issue or regulate standard activity methods and units for commercial enzymes (*11-12*). Marquardt and Bedford (*13*) reported that a similar problem occurred in the animal feeds industry. The lack of a standard method meant that the activity of commercial dextranases were being quoted by suppliers/vendors in many units, which only confused the factory buyer/user, and direct comparison of activities at the factory was not possible. Furthermore, the commercial dextranase market is very dynamic and activities and prices change regularly. To solve this, Eggleston (*14*) identified and modified a simple titration method (Figure 3) to measure the activity of dextranases in DU/mL units, which is now being successfully used by several U.S. factories.

Figure 3. Photograph of the use of the Eggleston (14) titration method to measure the activity of commercial dextranases at the factory (See page 2 of color inserts.)

The titration method (*11,14*) is easy to use, and there are no requirements for sophisticated equipment or a standard curve. Although the titration method measures the activity of dextranases under more optimal conditions than factory juice applications, it is highly correlated to a spectrophotmetric method (*1*), and higher industrial temperatures made no relative difference (*1*). Moreover, Eggleston and Monge (*1*) reported that IC-IPAD (ion chromatography with integrated pulsed amperometric detection) profiles of dextran/dextranase mixtures, used as a reference method, confirmed the accuracy of the relative activities of commercial dextranases by the titration method.

The urgent need for a standard method to measure the activity of dextranases at the factory or refinery was highlighted by the measurement of wide variations (up to 20-fold) in activity of commercial dextranases in the U.S. sugar industry that do not always reflect the costs of the enzymes (*11*). In some factories this has greatly reduced the dextranase efficiency because the users unknowingly added an dextranase of low activity. This is discussed further in the next section.

Differences in the Activities of Commercial Dextranases

Most commercial dextranases in the U.S. are produced from *Chaetomium gracile* or *erraticum* fungi, generally recognized as safe (GRAS), and formulated as liquids. Table I lists the activities, measured using the Eggleston factory titration method (*14*), of some dextranases commercially available in the U.S. and used in U.S. sugarcane factories.

Table I. Relative Activities of Dextranases in 2003 and 2004. From Eggleston *et al.* (*15*)

Commercial Dextranase	*Dextranase activity DU/mL*		*Classification*
	2003[a]	*2004*	
A	52,000	51,920	"concentrated"
B	57,687	ND[b]	"concentrated"
C	5,499	3,500	"non-concentrated"
D	4,786	2,750	"non-concentrated"

[a] From Eggelston and Monge (*1*)
[b] ND=not determined

U.S. dextranases occur in a wide range of activity, that Eggleston *et al.* (*1, 11, 15*) classified into "non-concentrated" (<25,000 DU/mL but usually <6,000 DU/mL) or "concentrated" (25,000-58,000 DU/mL but usually between 48,000-58,000 DU/mL) forms (Table I). An approximate 8-10 fold difference in activity existed between the two forms in 2003, and up to ~20-fold different in

2004 (Table I). Variation existed within each form as well. Similar large differences have also been reported in the activity of commercial dextranases available in South Africa (*16*). An approximate 9-fold difference exists in the activity of commercial α-amylases used in the sugar industry to hydrolyze starch, that also do not reflect the comparative unit costs of the enzymes (*17*).

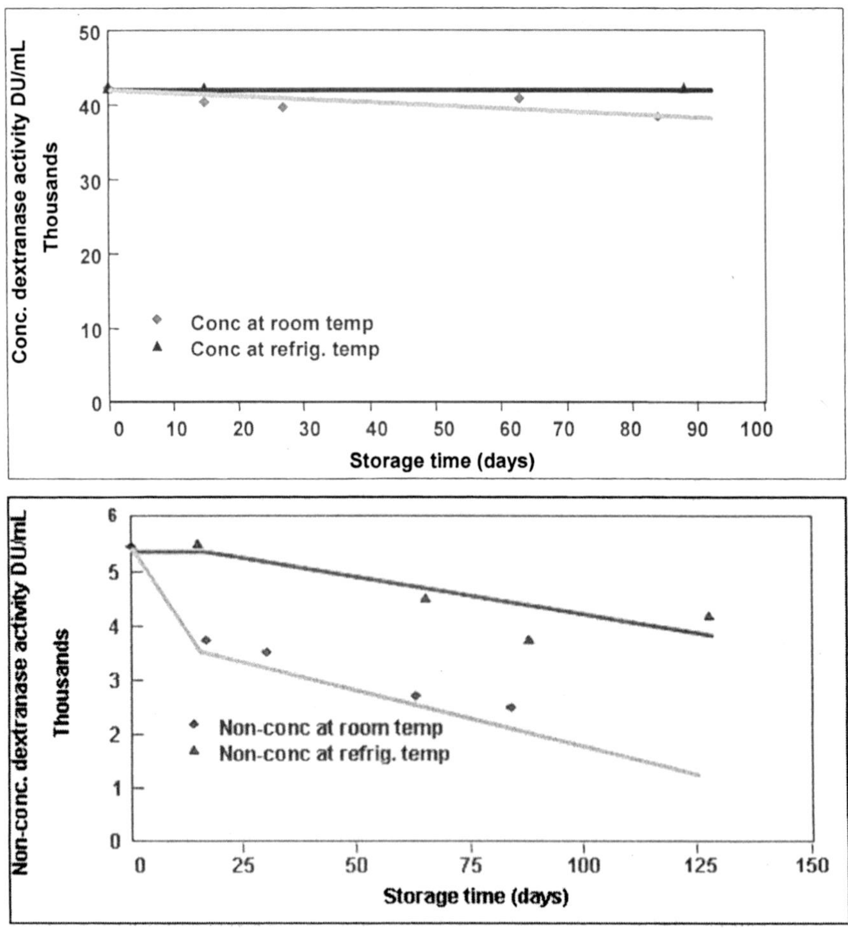

Figure 4. Changes in activity of "concentrated" and "non-concentrated" dextranases stored under (top) simulated factory storage conditions (ambient temperature ~25 °C), and (bottom) refrigerated conditions (4 °C) over a 90 day sugarcane processing season. From Eggleston and Monge (1).
(See page 3 of color inserts.)

Eggleston *et al.* (*1,15*) stated there are three important reasons to measure the dextranase activity at the factory/refinery: (i) to compare the economically equivalent activities of different commercial dextranases as vendor dextranase units differ, and activities and prices change regularly, (ii) to monitor the changing activities of dextranases on factory storage, and (iii) to measure the activity of delivered batches. The importance for reason (i) is highlighted in Table I, which shows that the "non-concentrated" dextranases were being sold with lower activities in 2004 than in 2003, at only a slightly lower price. The monitoring of activity on storage is extremely important, as the storage characteristics of commercial dextranases vary widely (*1*). Figure 4 illustrates, that under factory storage conditions (ambient temperatures 25 °C) over the typical length of a Louisiana processing season (~90 days), the activity of a "concentrated" dextranase decreased only slightly at ~9%, whereas the activity of a "non-concentrated" dextranase decreased by 50% and was not even stable when stored under refrigeration (Figure 4).

The dramatic loss in activity for the "non-concentrated" dextranase is because more water is available to de-activiate and denature the enzyme protein structure and increase its conformational mobility. Frost and Sullivan (*18*) reported that industrial enzymes have shorter shelf-lives than most industrial chemicals. Furthermore, because "non-concentrated" dextranases rapidly lose their activity, transport conditions can affect their activity, *e.g.*, they have arrived at the factory with no activity (*15*). Therefore, the activity of delivered batches to the factory should be monitored.

Industrial Conditions That Affect the Efficiency of Dextranases

The efficiency of dextranases in the factory depends on the pH, Brix (% dissolved solids) or water activity, temperature, retention time (R_t), agitation, and type, activity, and dosage of the dextranase applied (*7*). In many U.S. sugarcane factories there is insufficient retention time. Dextranase application to juice at the front (milling) end of the factory is limited by the size of juice tanks that often have < 6 min R_t. Dextranases cannot be applied at the clarification unit process because of the very high temperatures (~96 °C) and long R_ts up to 1.5 h) of clarified juice, and the lime added would have an inhibitory pH and ion effect on dextranases. Across the subsequent evaporation station, temperatures are also too high, although many factories have applied dextranase to syrup in last evaporators (*1*) where temperatures are ~65 °C and there is approximately 18 min R_t. Application of dextranases after the evaporator station is unwarranted because of the very low water activity/high Brix in massecuites (sucrose crystal/molasses mixtures). The optimum pH range for dextranase activity is pH

5.0-6.0, with the lower end of the range more preferable, which coincides with the typical pH operating range of juices before clarification in sugarcane factories.

Initial Laboratory Studies for Optimized Dextranase Factory Application

To understand how commercial dextranases react during industrial processing and how they can be manipulated, knowledge of how they performed under more ideal, laboratory conditions was first required (*19*). Therefore, a comprehensive laboratory study of the effect of dextranases on sugarcane juices and syrups was initially undertaken (*1,11*). From this study it was observed that both "non-concentrated" (5,999 DU/mL) and "concentrated" (52,000 DU/mL) dextranases at juice pH 5.4, showed similar maximum activity at ~50 °C (120 °F). The lowest activity occurred at 65.5 °C because of the partial denaturation of dextranase. This has dramatic consequences for the factory as most last evaporators have syrup temperatures ~65.5 °C. Dextranase activity was also low at 26-32 °C, typical ambient temperatures of juice, but still better than at 65.5 °C (*1*).

There was also a dramatic effect of Brix on the activity of dextranase (*1*). Dextranase activity was stable up to 25-30 Brix (Figure 5), but afterwards decreased rapidly because of the low concentration of water reactant. Overall, the pH, temperature and Brix conditions in factory last evaporators are sub-optimal for dextranase reactions. This was confirmed and shown to be uneconomical when both a "non-concentrated" (5,999 DU/mL) and a "concentrated" (52,000 DU/mL) were further added to final evaporator syrup in the laboratory (*1*). For every U.S. dollar spent on applying the "non-concentrated" dextranase to juice, $14.29 had to be spent for the equivalent dextran hydrolysis in syrup (*15*). This was only slightly better for the "concentrated" dextranase, where for every U.S. dollar spent on applying it to juice, $11.62 had to be spent for application to syrup.

Heating juice, with added "non-concentrated" or "concentrated" dextranase, to 50 °C dramatically hydrolyzed more dextran than at ambient juice temperatures (26-32 °C), and was much more economical (*1*). For "non-concentrated" dextranase (5,999 DU/mL), after 10 min at 10 ppm/juice and 50 °C, ~46.3% dextran was hydrolyzed compared to 13.6% at 32 °C. For "concentrated" dextranase (52,000 DU/mL), after 10 min at only 4 ppm/juice, 66.6% dextran was broken down at 50 °C (considered an over-dose), compared to 29.6% at 32 °C (*1*).

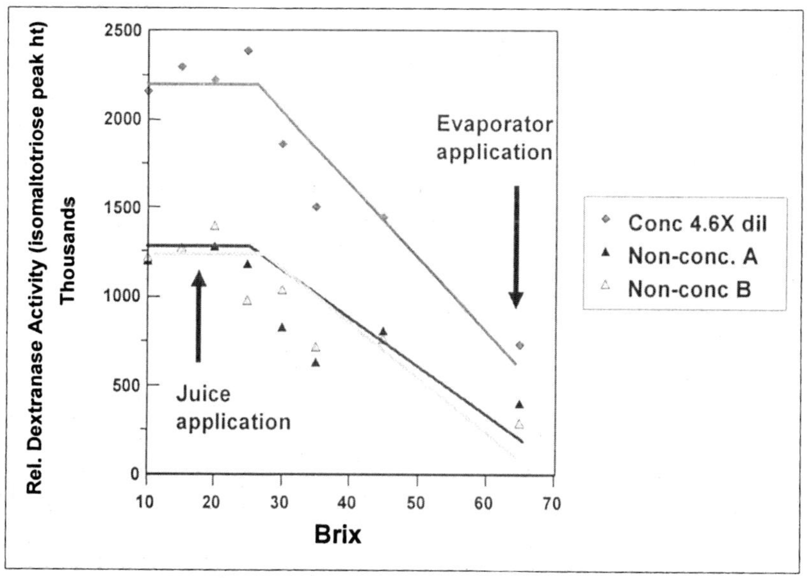

Figure 5. Effect of Brix on dextranase activity. The "concentrated" dextranase was diluted 4.6X to make it economically equivalent to the nearest priced "non-concentrated" dextranase. From Eggleston and Monge (1). *(See page 4 of color inserts.)*

Factory Trials for Optimized Dextranase Applications

If industrial processing conditions are ideal, an applied hydrolase enzyme such as dextranase will repeat the hydrolysis reaction many times during the process. However, the addition of hydrolases to an industrial process is often non-ideal, problematic, and difficult to optimize because conditions are frequently harsh. Also, extrapolation from laboratory and pilot plant scales to the industrial scale is not always linear. Although the initial laboratory results (*1*) gave a useful indication of the necessary dosage of "concentrated" dextranase to juice, they were obtained under ideal conditions compared to those at factories, where dextranase is added to much larger volumes of juice in tanks and pipes with fluctuating flow rates and agitation. Consequently, factory trials were conducted at two Louisiana factories across the 2004 sugarcane processing season. The laboratory results did, however, give a solid foundation from which to start the factory trials, and allowed the decision not to study dextranase applications to last evaporator syrups as they are not cost-effective (*15*).

Before we investigated the addition of a "concentrated" dextranase (52,000 DU/mL) to factory juice streams containing dextran, contact between the enzyme and substrate was considered. Smaller volumes will take longer to disperse in the juice tank. Figure 6 illustrates low contact between dextranase and dextran when only small volumes of a "concentrated" dextranase are applied. This was solved by applying working solutions of "concentrated" dextranase (*15*). Working solutions are prepared at the factory and represent the same final concentration of dextranase but at larger volumes to improve contact (Figure 6). Staff at factories are already used to preparing working solutions of flocculant chemicals for the clarification process. Moreover, applying working solutions of "concentrated" dextranase are much more cost effective than adding "non-concentrated" dextranase undiluted.

If working solutions of "concentrated" dextranases are to be useful, they need to be stable. Sucrose is a known stabilizer (*20*) of many industrial enzymes and, fortunately, is readily available at the factory. Sucrose occurs in its purest

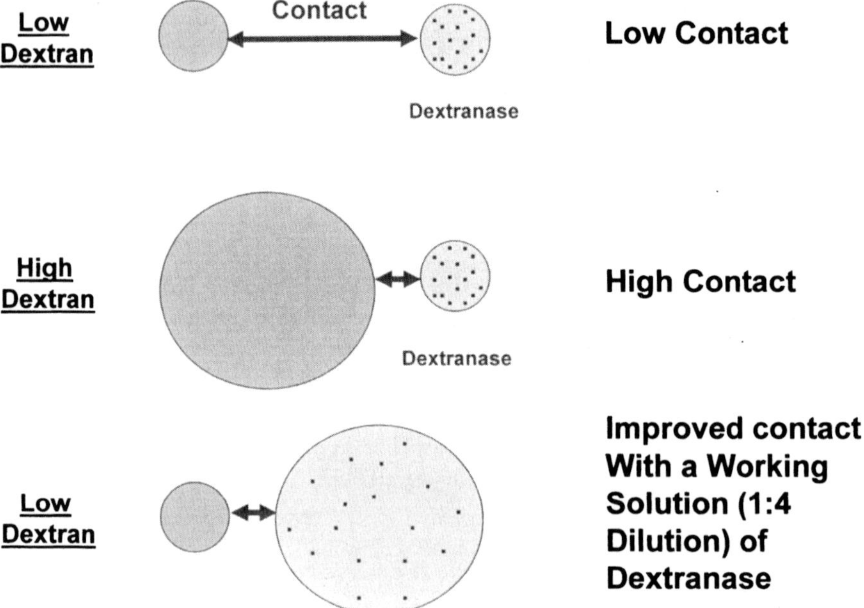

Figure 6. Diagram to illustrate the contact between dextranase and different concentrations of dextran. Circles depict volumes and squares depict enzyme molecules. The action of a working solution of "concentrated" dextranase (>25,000 – 58,000 DU/mL) to improve contact in factory process is also shown. Modified from Eggleston et al. (15).
(See page 4 of color inserts.)

form in raw sugar (~99.5% sucrose). A "concentrated" dextranase (52,000 DU/mL) diluted 5-fold with a 24 Brix raw sugar solution effectively stabilized the dextranase activity over 5 days, i.e., the activity decreased by only ~2% after ~140 h (*15*). Furthermore, "concentrated" dextranase diluted 2-fold with distilled or tap water, is still stable up to 48 h, and even 5-fold dilutions are stable for 24 h. Because the cheapest and most readily available source of water at the factory is tap water, it is conservatively recommended that a working solution be prepared with tap water and stored for 12 to 24 h maximum (*15*). If factory staff prefer, they can store the working solution for longer if it is prepared with a 24 Brix raw sugar solution.

The first Louisiana factory studied (*15*) had high antibody dextran (*21*) concentrations (>1,000 ppm/Brix) in juice. Applications of 2- or 5-fold working solutions (higher dosages than in laboratory studies were required at 6 ppm, that were normalized to the original enzyme activity) of "concentrated"dextranase (52,000 DU/mL) were successful in consistently hydrolyzing 70-94% dextran across a 5 min R_t tank. In strong comparison, the addition of 4.2 ppm (typical dosage applied previously by the factory staff) of "non-concentrated" dextranase (2,750 DU/mL) had neglibible effect on dextran hydrolysis in juice across the same 5 min R_t tank (*15*). At a second Louisiana factory, with lower antibody dextran concentrations (<300 ppm/Brix), addition of dextranase was more problematic (*4*). This was because large concentrations of dextran are easier to hydrolyze than low concentrations (see Figure 6), due to higher contact between the dextran and dextranase (low enzyme/substrate ratio). Below ~670 ppm/Brix dextran, dextranase efficiency decreased. Activity reduction accelerated below ~500 ppm/Brix antibody dextran (*4,15*). This can be solved if a working solution of higher dilution of "concentrated" dextranase is applied to improve contact, and if there is adequate mechanical agitation.

Most factory applications of dextranases to juice occur at ambient temperatures (26-32 °C) but Eggleston and Monge (*1*) recently showed that the maximum activity of many U.S. available dextranases in cane juice is ~50 °C. Therefore, heating juice at the factory may improve dextranase efficiency and, to some extent, overcome insufficient retention time. Although heating juice may contribute to a more optimum temperature for *Leuconostoc* growth and dextran formation, the addition of 10 ppm sodium carbamate biocide would inhibit this (*1*). A factory trial was conducted at a Louisiana factory to verify if heating juice under industrial conditions could improve dextranase application. As shown in Table II, just heating juice from 27 °C to 37 °C (the factory was unable to increase the juice temperature further) dramatically improved average dextran hydrolysis from 50.8 to 83.8%. Furthermore, this improvement generally occurred irrespective of the initial dextran concentration (Table II). This suggests that the efficiency improvement by heating juice occurs even when contact between the dextran and dextranase is sub-optimum. Although heating juice would be expected to increase factory energy inputs and costs, these will be negligible as existing heated juices could be recirculated into the juice tank or juice pipes. Moreover, because heating the juice will reduce the dextranase dosage, any costs from increased energy or biocide requirements will be significantly lower than costs for the relatively expensive dextranase.

Table II. The Effect of Heating Cane Juice on the Ability of a "Concentrated" Dextranase (52,000 DU/mL) to Hydrolyze Dextran at a Louisiana, U.S. Factory in the 2005 Processing Season. Conditions: 7.5 ppm Dosage; 5 Min Retention Time; 5-Fold Working Solution Prepared with Distilled Water.

Sample number[a]	*Antibody Dextran (ppm/Brix)*[b]		*% Dextran*
	$Tank_{IN}$	$Tank_{Out}$	*Hydrolysis*
	Juice Temperature ~27 °C		
1	5686	1159	79.6
2	5728	3248	43.3
3	6266	2920	53.4
4	3862	2188	43.4
5	4210	1174	72.1
6	4280	3723	13.0
Average:	**5005**	**2402**	**50.8**
	Juice Temperature ~37 °C[c]		
7	5985	1044	82.6
8	4942	2628	46.7
9	2169	175	91.9
10	3292	469	85.8
11	4748	846	98.2
12	5445	119	97.8
Average:	**4430**	**753**	**83.8**

[a] Samples were collected of juice entering (IN) and exiting (OUT) the juice tank, taking into account the 5 min retention time. Samples were collected every 7 min and biocide added immediately to prevent further degradation. Samples were frozen until analyzed.
[b] Dextran was measured using a monoclonal antibody method (*21*)
[c] Juice was heated by recirculating heated juice into the tank. Juice temperature was allowed to equilibriate at ~37 °C for 30 min before sampling.

As a result of this factory research, several factories in Louisiana and Florida (T. Johnson, personal communication) have now changed from using a "non-concentrated" dextranase to a "concentrated" dextranase, applying the latter as a working solution. Two Louisiana factories (D. Stewart, personal communication) are already heating the juice to improve dextranase applications, and more factories are expected to change soon. The authors recommend a little further experimentation in the factory to optimize dextranase applications under each factory's unique conditions.

Other Possible Short-Term Methods to Improve the Practical Industrial Application of Dextranases

As stated in the above section, to improve contact between the dextran and dextranase in heated juice, a working solution of high dilution of "concentrated" dextranase can be applied. Serpentine pipes would provide turbulent flow for better mixing and contact of the dextran and dextranase without the added costs of mixers, occupy relatively little space in the often crowded factory/refinery, and increase retention time (*4*).

An alternative method for industrial enzyme application is enzyme immobilization on a solid support. There are different methods to immobilize enzymes (*22*): covalent attachment to a solid support, entrapment in a gel matrix, adsorption onto an insoluble matrix, and intermolecular cross-linking of the enzyme to form an insoluble matrix. Advantages of immobilization are the re-use of the enzyme, continuous output of products, control of the reaction rate by regulating flow rate, easy stoppage of the reaction for the facile removal of the enzyme, non-contamination of the product, and increased enzyme stability. Disadvantages of immobilization are often reduced volumetric activity of the enzyme (*23*) and mass transfer problems. Immobilized enzyme systems generally require the substrate to be a relatively small molecule, i.e., glucose in high fructose corn syrup (HFCS) production, because the substrate and resulting product must be easily moved in and out of the immobilized support matrix. As dextran is a large molecule it would limit the pore size of the immobilized system used. Furthermore, immobilized enzyme systems are also expensive, particularly the initial capital cost. As processing seasons in the U.S. sugarcane industry are typically 3-4 months, cost issues are even greater.

A more promising technology to improve the industrial efficiency of dextranases is the use of low sonication microwave technology. Yachmenev *et al.* (*24-25*) revoked the common perception that shock microwaves resulting from the collapse of cavitation bubbles will severely damage and inactivate the very sensitive and intricate structure of the enzyme protein. In reality, low level and uniform ultrasonication microwaves are an ideal stirring mechanism (*24-25*) for the enzyme and substrate, improve contact, and increase the overall reaction rate. There is also improved removal of products of enzymatic hydrolysis from the reaction zone. However, this technology has not been commercialized yet (*24-25*).

Future Outlook

Typically, industrial enzymes are very sensitive to pH, temperature, water activity, and contaminants. Therefore, it is very difficult to improve the operating conditions of enzymes to optimize their functional abilities. Industrial enzymes are also difficult to store since their shelf-life is shorter than most industrial chemicals. This chapter highlights the practical difficulties in applying dextranases to hydrolyze dextran in sugar manufacture. Dextranase is often referred to as a "band-aid" enzyme in the enzyme industry because it is only applied when there is a processing problem, *i.e.*, in the sugar industry dextranase is applied when there has been sugarcane or sugarbeet deterioration (or when deteriorations is expected because adverse weather occurred). Consequently, dextranases only have a relatively small market and low volume sales compared to other industrial enzymes. Thus, very limited research and development by large enzyme companies has been undertaken to tailor the dextranases' properties to the harsh processing conditions. For example, better performance at the high Brix/low water conditions of syrup in last evaporators is required. As a consequence, optimization studies to improve the operating conditions of dextranases as outlined in this book chapter are the best solutions for the short-term. The use of serpentine pipes to increase mixing and reduce the need for more retention time in the factory/refinery, and promising low level, uniform ultrasound technology (*24-25*) will only enhance industrial optimization.

More long term solutions to overcome the processing properties and storage constraints of dextranases could be protein engineering of the enzymes. Protein engineering techniques include site-directed mutagenesis and random mutagenesis (directed evolution) (*26*). Further genetic engineering of the fungal source of the dextranases would improve production yields and purification (*12*).

Acknowledgements

The authors thank the American Sugar Cane League for contributing funds to this research. Mention of trade names or commercial products in this article is solely for the purpose of providing specific information and does not imply recommendation or endorsement by the U.S. Department of Agriculture.

References

1. Eggleston, G.; Monge, A. *Process Biochem.* **2005**, *40*, 1881-1894.
2. De Bruijn, J. M. *Zuckerindustrie* **2002**, *125*, 898-902.
3. Khalikova, E.; Susi, P.; Korpela, T. *Microbiol. Mol. Biol. Rev.* **2005**, *69(2)*, 306-324.

4. Eggleston, G.; Monge, A.; Montes, B.; Stewart, D. *Int. Sugar J.* **2006b**, in press.
5. Tilbury, R. H. *Proc. Int. Soc. Sugar Cane Technols.* 14th Congress, **1971**, 1444-1458.
6. Hidi, P.; Staker, R. *Proc. 42nd Conf of Queensland Society Sugar Cane Technols.* **1975**, 331-344.
7. Inkerman, P.A. *Proc. Int. Soc. Sugar Cane Technols.* **1980**, *17*, 2411-2427
8. Polack, J. A.; Birkett, H. S. *J. Am. Soc. Sugar Cane Technols.* **1978**, *43*, 307-315.
9. DeStefano, R. P. *J. Am. Soc. Sugar Cane Technols.* **1988**, *8*, 99-104.
10. Edye, L. A.; Clarke, M. A.; Kitchar, J. *Proc. Sugar Indust. Technols.* **1997**, *56*, 359-362.
11. Eggleston, G.; Monge, A. *Proc. Sugar Proc. Res. Conf.* **2004**, p 371-394.
12. Eggleston, G. In *Industrial Application of Enzymes on Carbohydrate Based Materials*; Eggleston, G.; Vercellotti, J.R.; Eds.; ACS Symp. Series: Washington D.C. **2007**, Chapter 1, in press
13. Marquardt, R. R.; Bedford, M. In *Enzymes in Poultry and Swine Nutrition*, Marquardt, R. R.; Zhengkang, H.; Eds.; **1997**, IDRC publication: Canada, p 154.
14. Eggleston, G. *Sugar J.* **2004**, *67*, 32-33.
15. Eggleston, G.; Monge, A.; Montes, B.; Stewart, D. *Int. Sugar J.* **2006a**, *108(1293).*
16. Anon. SMRI annual report, *Sugar Milling Research Institute publication*, **2004**, Durban: South Africa, p 18.
17. Eggleston, G.; Monge, A., Montes, B.; Guidry, D. **2006c**. *Process Biochem.* **2007**, in press.
18. Frost and Sullivan. *Laboratory talk*, http://www.laboratorytalk.com/news/fro/fro189.html, **2004**.
19. Tucker, G.A. In *Enzymes in Food Processing*, Tucker, G.A.; Woods, L.F. J.; Eds.; Blackie, **1995**, p 1-25.
20. Davidson, P. S. *Pharm. Res.* **2001**, *18(4)*, 474-479.
21. Rauh, J.S.; Cuddihy, Jr., J.A.; Falgout, R.N. *Proc. Sugar Industry Technologists 58th Annual Meeting*, **1999**, p 17-27.
22. Worsfield, P. J. *Pure Appl. Chem.* **1995**, *67(4)*, 597-600.
23. Lievonen, J. *Technical Paper of the VTT (Technical Res. Center of Finland),* **1999**, *No. 43/99*, 22-36.
24. Yachmenev, V.; Blanchard, E.; Lambert, A. *Ultrasonics* **2004**, *42*, 87-91.
25. Yachmenev, V.; Lambert, A. In *Industrial Application of Enzymes on Carbohydrate Based Materials*; Eggleston, G.; Vercellotti, J.R.; Eds.; ACS Symp. Series: Washington D.C. **2007**, in press
26. Hashida, M.; Bisgaard-Frantzen, H. *Trends Glycosci. Glyco.* **2000**, *12*, 389-401.

Chapter 7

Effect of Commercial Enzymes on Color and Total Polysaccharide Content in Sugarcane and Sugarbeet Juice

Marianne McKee, Mary An Godshall, Sara Moore, and Ron Triche

Sugar Processing Research Institute, Inc., 1100 Robert E. Lee Boulevard, New Orleans, LA 70124

In sugar processing, low color is an important quality requirement for sugar manufacturers and consumers. Sucrose produced from either sugarcane or sugarbeet is first extracted as a raw juice that has high color and polysaccharide content. Many processing steps are required to clarify and decolorize the solution to obtain the final white sugar product. The effect of targeted commercial enzymes was examined on reducing color and polysaccharide concentration in raw cane and beet juices. Juices were treated with 28 commercially available enzymes (500 ppm enzyme concentration for 30 min at 50 °C) and monitored for color and total polysaccharide content. For sugarcane juice, enzymes with hemicellulase, cellulase, xylanase, and glucosidase activity were the most effective in removing color and/or polysaccharides. Hemicellulase, pectinase, xylanase, and glucanase removed color and/or polysaccharides from sugarbeet raw juice.

Introduction

Sources of Sucrose

There are two main sources of sucrose (commonly known as table sugar): sugarcane and sugarbeet. Sugarcane is primarily grown in warmer tropical climates such as the southern United States and South America. Sugarbeets are more cold tolerant and grown in colder climates such as Europe and the northern United States. In the sugarcane plant, most sucrose is found in the stalk, while in sugarbeets, sucrose is found in the root system below ground. Although these two sources are vastly different, the end product from each is the same – crystalline sucrose (β-D-fructofuranosyl-α-D-glucopyranoside). The purification process for each system is different, but the initial juice solution that sucrose must be extracted from is dark brown in color, and filled with many impurity components other than sucrose such as polysaccharides, colorant molecules, and amino acids for both sugarcane and sugarbeet systems.

For the sugarcane system, the sugarcane stalks are harvested from the fields and transported to the factory, where they are crushed to extract the juice containing the sucrose. This juice is referred to as raw juice. The juice then undergoes hot lime clarification to remove polysaccharides, turbidity, and other material from the raw juice solution. This clarified juice is then concentrated and boiled (evaporated) to crystallize a raw sugar product, with molasses formed as a side product. This raw sugar can range in color from light to dark brown. The raw sugar is shipped to a sugar refinery for further color removal using either ion exchange resins, bone char, or other techniques to obtain the white sugar product.

The sugarbeet system is different. Sugarbeets are harvested and transported to the factory where they are sliced to extract the raw juice containing sucrose. This raw juice undergoes carbonatation clarification, and concentration to form the solution from which the sugar is crystallized. White crystalline sucrose can be crystallized directly from beet thick juice solution. This is compared to the sugarcane system where a highly colored intermediate raw sugar product must be further refined to obtain the white sugar product.

Currently the sugar industry employs two enzymes to aid the processing of sugarcane or sugarbeets to raw or white sugars. α-Amylase is used to hydrolyze and remove starch (α-(1→4)-α-D-glucan) or amylose from solution. High starch content in the raw or clarified juice can lead to many processing problems such as slow boiling rates leading to lower sucrose yields. Another problem in sugar processing that is currently being addressed by enzymes, is unwanted dextran (α-(1→6)-α-D-glucan) that is formed when sugarcane or sugarbeet deteriorates. Dextranase is used to hydrolyze and remove dextran from raw juice solutions for easier processing, as discussed in another chapter of this book (*1*).

Sugar Color

Sucrose is colorless; it is the non-sugar components of the crystal that give sucrose its color. White sugar or very low color sugar is preferred by both industrial customers and household consumers. Sugar color can vary from dark brown, almost black as in molasses, to a light golden brown to pale yellow and even to white or what appears to be colorless. Because the color of sugar solutions can vary so widely, it is very important to have a standardized system to measure and compare the color of sugars. The International Commission for Uniform Methods of Sugar Analysis (ICUMSA) publishes standard methods for analysis of sugar products in trade. The method for color analysis (*2*) requires measuring the absorbance of the sugar product at 420 nm and pH 7 after filtration through a 0.45 μm membrane. The resulting color is reported in ICUMSA units (IU) to allow worldwide standard comparison of sugar product color.

There are two main sources of sugar color: (i) plant derived or (ii) formed during processing. In sugarcane processing, the colorants tend to be natural plant pigments associated with polysaccharides found in the sugarcane plant. These colorants change very little during processing (*3*). In comparison, sugarbeet colorants are generally produced during the processing of the sugarbeets. These colorants are mostly alkaline degradation products of fructose and glucose which, in turn, are products of the acid or enzymatic degradation of sucrose. Another type of colorant commonly found in sugarbeet processing is melanoidins (*4*). These colorants are also formed by the reaction of sucrose with materials in the raw juice solution during the processing of the sugarbeets.

There are four main types of sugar colorants. The first type of colorant is phenolics. The precursors for phenolics found in the sugarcane plant are flavonoids and cinnamic acid derivatives. As the color intensity of these colorants is pH sensitive, the pH of the processing solution can greatly affect the color when this type of colorant is present.

A second type of colorant is the melanoidins. These colorants are formed by Maillard or browning reactions. These reactions occur between amino acids and reducing sugars present in the sugar processing solution. The resulting sugar solution product is usually very dark.

Caramels are the third type of colorants commonly found in sugar processing. This class of colorant is formed by the thermal degradation of sucrose and other carbohydrates into a complex reaction mixture. These colorants can range in color from yellow to brown to very dark black, depending on the amount of degradation that occurs.

The fourth group of colorant molecules is referred to as invert (glucose and fructose) degradation products from the reactions of sucrose degradation products with other components in the sugar processing solutions. High pH causes rapid degradation and darkening of the sugar solution. Degradation can also occur at low pH, but it is much faster at high pH.

Sugarbeet processing products have lower color than the sugarcane processing products. White beet sugar can be crystallized from a much darker thick juice solution than the white cane sugar. Therefore, the raw sugar produced from sugarcane must be further refined to remove more color to produce "whiter" sugar for sale to the consumers. One reason sugarcane contains more color is in part due to the source of sugarcane colorants being mainly the sugarcane plant itself. Figure 1 below shows the relative color of the sugarcane and sugarbeet products. White sugar color is typically in the range of 20 IU. As shown in Figure 1, cane sugar processing products color range from over 14,000 IU for the raw juice to 1000 IU for the raw sugar, and down to white sugar levels after refining. The most effective color removal step for the sugar is crystallization. In sugar processing, more color is transfer to the crystal in the sugarcane than sugarbeet systems. If enzymes can be used to reduce the color of the sugar processing products by degradation of color forming components in raw juice, less color may be transferred to the crystal in both the sugarcane and sugarbeet systems.

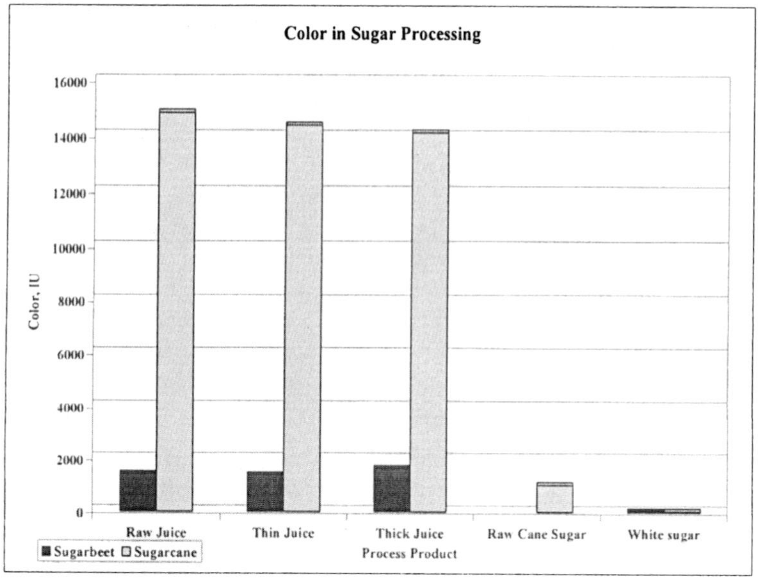

Figure 1. Color in sugar processing products.

Low sugar color is important because it is generally used as an indicator of quality. It is usually perceived that the lower the color, the higher the quality of the sugar. Many times industrial users require extremely white sugar – for example, in the manufacture of clear beverages and carbonated beverages. In some countries, such as South Africa, there are penalties for producing sugar that does not meet the required color quality standards. Currently in the United States no such penalties exist for sugar color, but penalties are in place for other quality indicators such as the dextran content of the raw sugar delivered to the refinery.

Polysaccharides in Sugar Processing

Another problem in sugar processing is polysaccharides. Processing problems caused by polysaccharides can lead to delays and expensive sugar losses in factories and refineries. Polysaccharides decrease the filterability of solutions in factories and refineries that can cause major delays and cost time and money. Another effect of these polysaccharides is that boiling or crystallization rates are slow, leading to lower sucrose yields and profits for the factories and refineries. In the crystals that do form, the crystal structure can be elongated, leading to problems in centrifugation, and ethe processing of end user products. For example, when sucrose with elongated crystal structure is used in hard candy production, the candy does not form as expected leading to problems for candy producers (*5*) .

As with colorants, sugarcane and sugarbeets have different polysaccharides. For the sugarcane system, the common polysaccharides are those found in the sugarcane plant, such as the indigenous sugarcane polysaccharide (*6*). This polysaccharide is an arabinogalactan found in soluble cell wall hemicellulose. Another common polysaccharide in sugarcane processing is dextran (α-(1→6)-α-D-glucan), which also can occur in sugarbeet processing. Dextran is a microbial polysaccharide that results from the infection of the sugarcane mainly with *Leuconostoc* bacteria, after physical damage to the sugarcane plant in the field or a delay in processing the cane after harvest. This bacterium utilizes sucrose to produce dextran. Another detrimental polysaccharide is starch (α-D-(1→4)-glucan) found mostly in the leaves and growing points of the sugarcane plant. This starch is transferred to the raw juice solution when the sugarcane is crushed in the mill.

The main polysaccharide found in sugarbeet processing is pectin. The majority of the pectin structure consists of partially methylated poly-α-(α-(1→4)-D-galacturonic acid residues with areas of alternating α-(1→2)-L-rhamnosyl-α-(1→4)-D-galacturonosyl sections containing branch-points with mostly neutral side chains of mainly L-arabinose and D-galactose. This polysaccharide can be found in relatively high concentration of 1-2% on beets. Another polysaccharide found in sugarbeets (*7*) is a galactan and araban that has not been well defined and probably represents a soluble hemicellulose.

The study reported here has short-term and long-term goals that may impact sugar processing. The short-term goal was to determine if certain enzyme functionalities could reduce color and polysaccharides of sugar processing products in either sugarbeet or sugarcane systems. The long-term goals for sugar factories include using the enzymes to aid in filtration by removing polysaccharides, and slowing or preventing color formation in process samples.

Experimental

The experimental protocol for the sugarcane or sugarbeet raw juice samples included using raw juice collected from sugar factories with the same enzyme treatment procedure. The experiments were not optimized for the enzyme reaction conditions, but conducted at conditions of raw juice processing found n either sugarcane or sugarbeet factories. Enzyme concentration of 500 ppm dosage was added to juice (100 mL) and the mixture allowed to react at 50 °C for 30 min. The native pH of the raw juice, which is usually 5.5 to 6.0, was used. After treatment with the enzyme, the solution was tested for color, total polysaccharides, and sucrose degradation. After treatment with enzymes the raw juice solutions were analyzed by ICUMSA methods for color and turbidity. The SPRI total polysaccharide method was used (*8*), which is widely used in the sugar industry. Twenty-eight commercially available enzymes were tested on the juice samples. The enzyme functionalities that were tested included cellulase, xylanase, pectinase, amylase, glucanase, glucosidase, polyphenol oxidase, hemicellulase, pullulanase, and arabanase. Table I shows the components and functionalities of the commercially available enzyme mixtures tested. A control sample for the sugarcane and sugarbeet raw juice was prepared by heating 100 mL juice to 50 °C for 30 min.

Results and Discussion

Control Sample

Previous work at SPRI had shown that heating raw juice slightly reduces color, polysaccharide content, and turbidity of the juice sample because of natural floc formation from the coagulation of colloids that precipitate out and removes these components. Therefore, a control sample of heated sugarcane or sugarbeet juice was prepared for each set of experiments. For the sugarcane experiments, the heated control sample removed 12.2 to 13.2% color and 10.9 to 12.2% total polysaccharides. Also, heating the juice increased the turbidity 11.0

Table 1. Enzyme Functionalities

Commercial Enzyme	*Functionality*
AMG 300L	Amyloglucosidase
BAN 240L	Amylase
Cellulase	Hydrolyzes cellulose
Celluclast	Cellulase and other carbohydrases
DEPOL 40L	Cellulase, xylanase, and pectinase
DEPOL 112L	Cellulase, xylanase
DEPOL 670L	Cellulase, xylanase, pectinase, ferulic acid esterase
Driselase	Cellulase, pectinase, laminarinase, xylanase, and amylase
Fermcolase	Oxidase
Finizyme	β-Glucanase
α-glucosidase	Hydrolyzes terminal, non-reducing 1,4 linked α-glucose residues
β-glucosidase	Hydrolyzes terminal, non-reducing 1.4-linked β-glucose residues
Hemicellulase	Liberates galactose form hemicellulose
Inulinase	1-β-D-fructan-fructanhydrolase
Laccase	Polyphenol oxidase
LE-R	Lysing enzyme from *Rhizoctonia solani* containing yeast glucanase activity
LE-T	Lysing enzyme from *Trichoderma harzianum*, containing cellulase, protease, and chitinase
Pectinex 3XL	Pectinase
Pectinex Ultra	Pectinase
Pullulanase	Hydrolyzes (1→6)-α-D-glucosidic linnkages in pullulan
Rapidase A	Pecitnase and hemicellulase
Rapidase P	Amylase and pectinase
Rapidase ExColor	Pecitanse with hemicellulase
Rapidase X-Press	Pectolytic enzyme
Sucrodex	Dextranase, amylase, cellulase, β-glucanase, and xylanase
Viscozyme 120L	"carbohydrase"
Viscozyme	Arabanase, cellulase, β-glucanase, pectinase, xylanase
Xylanase	Hydrolyzes xylan

to 12.5% compared to unheated juice. The color and polysaccharide content of the heated control samples were used as the baseline measurement to compare with the enzyme treated samples.

Sugarcane Enzyme Treated Samples

Raw juice sugarcane samples were analyzed for color and polysaccharide content. A decrease in color or total polysaccharide content was considered significant when color or total polysaccharide of the sample was 10% or more lower than the control sample. Nine of the 28 enzymes tested (Table I) were found to remove a significant amount of color. The nine enzymes that removed ten percent or more color from the raw mixed juice for the sugarcane systems were: Celluclast, DEPOL 40L, DEPOL 112L, DEPOL 670L, Driselase, α-glucosidase, β-glucosidase, hemicellulase, and Viscozyme L. Hemicellulase removed the most color at 18% and Driselase removed ~12% color with the other enzymes listed falling within this range. These findings are summarized in Figure 2. The exact structure of the colorant molecules in the raw sugarcane juice is not well defined, but the enzymes listed above were found to break specific bonds within the structure of the colorants in order to reduce the overall color of the raw juice solution. The functionalities that reduced color included cellulase, xylanase, hemicellulase, and glucosidase.

Total polysaccharide analysis showed that eight enzymes removed ten percent or more of the polysaccharides from the sugarcane raw juice samples. These eight enzymes were: DEPOL 40L, DEPOL 670L, Finizyme, Inulinase, Pectinex 3XL, Rapidase P, Viscozyme 120L and Viscozyme L. Viscozyme L removed ~16% total polysaccharide and Finizyme removed ~11% with the other enzymes falling within this range. These results are summarized in Figure 2. These enzymes were effective at breaking down the polysaccharides found in sugarcane raw juice samples. These include the indigenous sugarcane polysaccharide mentioned in the Introduction Section. Structure and exact linkage information about this polysaccharide are not well defined. The enzyme functionalities that were effective at reducing polysaccharides in sugarcane raw juice are cellulase, xylanase, pectinase, glucanase, hemicellulase, and xylanase. These enzyme functionalities were either alone or in an enzyme mixture that may have led to the breakdown of polysaccharide molecules.

However some enzymes tested increased the polysaccharide content of the juice. These enzymes included hemicellulase, Driselase, Laccase, and xylanase. These were analyzed for the presence of starch as they were solid formulations and may have been stabilized with starch for storage. These enzyme preparations were found to contain starch. This could be a potential problem with laboratory scale experiments, but may not affect factory applications.

Current enzyme applications in the sugar industry, *i.e.*, α-amylase and dextranase, are liquid formulations. Research is being undertaken to formulate these enzymes as liquids.

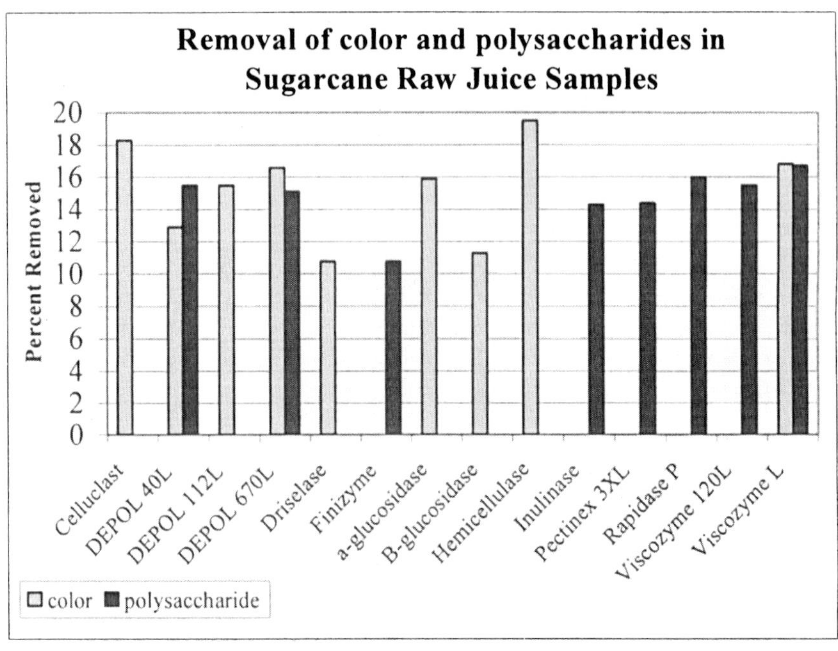

Figure 2. Results of color and polysaccharide removal by commercial enzymes in raw sugarcane juice. Only the enzymes that removed 10% or more color and/or polysaccharides are shown.

One of the most important requirements for the use of enzymes in sugar processing is that the enzyme does not degrade expensive sucrose. The raw juice solutions treated with enzymes that removed ten percent or more color or polysaccharides were analyzed for sucrose degradation. This analysis was undertaken by ion chromatography with pulsed amperometric detection, monitoring the formation of glucose and fructose. The enzymes that showed the production of sucrose degradation products were DEPOL 40L, DEPOL 670L, α-Glucosidase, Pectinex 3XL, and Viscozyme L. These enzymes cleaved the α(1→2) glycosidic bond between glucose and fructose, destroying the sucrose molecule and, therefore, would not be suitable for use in sugar processing. Enzymes that did not degrade sucrose were cellulase, Cellclast, DEPOL 112L, Driselase, Finizyme, β-glucosidase, hemicellulase, and xylanase, and would potentially be safe to use in sugar processing with respect to sucrose degradation.

One exception to this above observation may be Viscozyme. Viscozyme removed significant amounts of color (16.8%) and polysaccharides (16.7%) but produced glucose, a degradation product of sucrose. Glucose hydrolysis from the terminal end of polysaccharides is a function of the Viscozyme enzyme mixture. No increase of fructose was observed in the sucrose analysis chromatograms for the viscozyme treated samples, indicating the increase in glucose was not from the degradation of sucrose, but rather from the hydrolysis of polysaccharides. Further investigation is needed before using this enzyme in sugar processing to determine if Viscozyme does degrade sucrose under these reaction conditions.

Summary of Sugarcane Juices Treated with Enzymes

The three enzyme functionalities that significantly reduced color in the raw juice were cellulase, xylanase, and hemicellulase. Cellulase also decreased polysaccharide content in sugarcane juice. Hemicellulase and xylanase increased polysaccharide content in juice possibly due to starch stabilizer in the enzyme formulation. Commercial mixtures of enzymes also proved useful in reducing color or polysaccharides in sugarcane juices. These mixtures included DEPOL 112L a mixture of cellulase and xylanase that removed 15% color and 9.5% polysaccharides. Another mixture that proved useful was DEPOL 670L. This mixture of cellulase, xylanase, pectinase, and ferulic acid esterase removed a significant amount of color and polysaccharides, but, unfortunately, degraded sucrose.

Sugarbeet Enzyme Treated Samples

The control sugarbeet sample was prepared the same as the sugarcane control sample. Diffusion juice (100 mL) was heated to 50 °C for 30 min. Heating the diffusion juice removed 9% of the total polysaccharides. Heating the diffusion juice increased the color by 2% and also increased the turbidity by 2.6%.

As with the sugarcane samples, sugarbeet samples were tested for color and polysaccharide removal. A decrease in color or total polysaccharide content was considered significant when color value or total polysaccharide content for the sample was 10% or more below that of the control sample. Only four of the 28 enzymes tested lowered color by ten percent or more: Fermcolase, Finizyme, hemicellulase, and xylanase. Xylanase removed the most color at 31.4 % and the Fermcolase removed ~17% color with the other enzymes falling within this range. These findings are summarized in Figure 3.

Four enzymes removed ten percent or more of the total polysaccharides from the raw sugarbeet solutions: DEPOL 40L, DEPOL 670L, Pectinase Ultra, and Viscozyme L. Viscozyme L removed ~19.9% total polysaccharide and DEPOL 40L removed ~14.8% with the other enzymes falling within this range. These results are summarized in Figure 3.

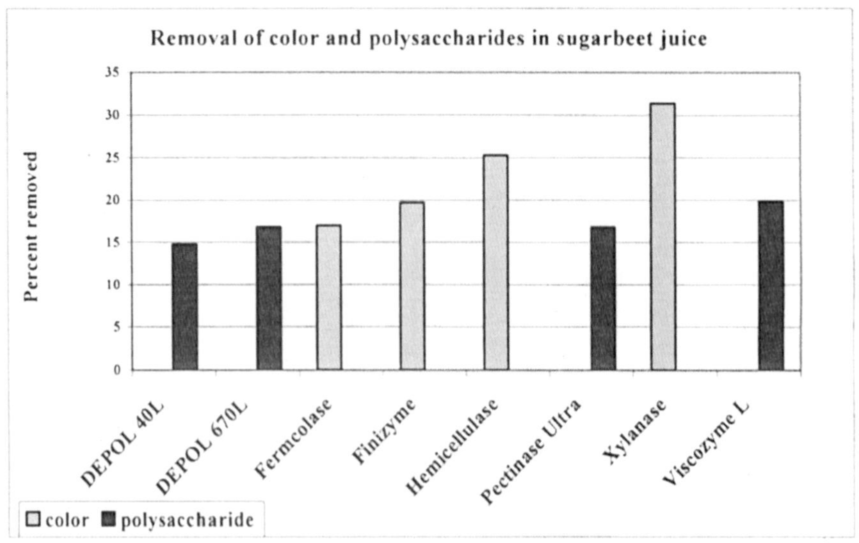

Figure 3. Results of color and polysaccharide removal by commercial enzymes in raw sugarbeet juice. Only the enzymes that removed 10% or more color and/or polysaccharides are shown.

Similar to the sugarcane samples, some enzymes increased the polysaccharide content of the juice. These were hemicellulase and xylanase that may, unfortunately, contain starch as a stabilizer.

The raw juice solutions of the enzymes that removed ten percent or more color or polysaccharides were analyzed for sucrose degradation by ion-chromatography. The enzymes that degraded sucrose were DEPOL 40L, DEPOL 670L, Pectinex Ultra, and Viscozyme L, and are not be suitable for use in sugar processing. Enzymes that did not degrade sucrose were Finizyme, hemicellulase, and xylanase. As discussed with regard to the sugarcane processing, one exception to the above observations may be Viscozyme.

Summary of Sugarbeet Sample Results

The four enzymes that significantly reduced color in the raw juice were Fermcolase, Finizyme, hemicellulase, and xylanase. Hemicellulase and xylanase increased polysaccharide content in juice possibly due to polysaccharide stabilizers in enzyme formulation. Commercial mixtures of enzymes also proved useful in reducing color or polysaccharides in sugarcane juices. DEPOL 40L and DEPOL 670L are two commercial enzyme mixtures that reduced polysaccharide content in sugarbeet juice. DEPOL 40L is a mixture of cellulase, xylanase, and pectinase. Cellulase, xylanase, pectinase, and ferulic acid esterase combine to make DEPOL 670L.

Conclusions

Twenty-eight commercially available enzymes or enzyme mixtures were tested for reactivity in sugarcane and sugarbeet raw juices for removal of color and/or polysaccharides. The enzyme mixtures that removed a significant amount (greater than ten percent) of either color or polysaccharides were then tested for sucrose degradation. These reactions were conducted under sugar processing conditions. It is important to note that these reactions were not conducted under optimized enzyme reaction conditions and, therefore, not at the optimum for the enzyme being tested. This was a preliminary study into the use of enzymes for the removal of color and polysaccharides from raw juice solutions to determine if enzymes would be a useful tool in removal of these components. Requirements for enzyme use in commercial sucrose processing include the following: (i) It must be commercially available, (ii) approved for food processing (GRAS - generally recognized as safe by the Food and Drug Administration), (iii) low cost, and (iv) not degrade sucrose. These preliminary experiments have shown that the enzymes studied may be more useful in sugarcane processing rather than in sugarbeet processing. More enzymes were found to remove the types of colorants and polysaccharides found in sugarcane rather than those found in sugarbeet. Also, processing conditions for sugarcane may be more optimum for enzyme activity because lower temperature and pH are used in sugarcane processing compared to sugarbeet processing. With more investigation based on the findings in this work, more specific enzymes may be found for the sugarcane system while more active enzymes may be found for the sugarbeet system.

By removing colorants and polysaccharides from raw juice solutions, sugar factories may not only produce a lower color sugar, they may also see improvement in boiling or crystallization. This would be achieved by removing polysaccharides that can slow sucrose crystallization. Ultimately, the use of new enzymes in sugar processing may improve sucrose yields as well as produce a lower color or higher quality product for sugar factories, leading to an increase in profit for sugar manufacturers.

References

1. Eggleston, G.; Monge, A.; Montes, B.; Stewart, D. In *Industrial Application of Enzymes on Carbohydrate Based Materials*, Eggleston, G., Vercellotti, J.R.; Eds.; ACS Symposium Series*: Washington D.C.* **2007**, in press.
2. ICUMSA Methods Book GS1/3-7, **2002**, p 1-3.
3. Godshall, M.A.; Vercellotti, J.R.; Triche, R.D. *Int. Sugar J.* **2002**, *104*, 228-223.
4. Godshall, M.A.; Chou, C.C. *Proc. Sugar Res. Proc. Res. Conf.* **2000**, p 122.
5. Haynes, L; Zhou, N.; Hopkins, W. *Proc. Sugar Res. Proc. Res. Conf*, **2004**, p 138.
6. Roberts, E.J.; Godshall, M.A.; Carpenter, F.G.; Clarke, M.A. *Int. Sugar J.* **1976**, *78*, 163-165.
7. Clarke, M.A; Roberts, E.J.; Godshall, M.A.; deBrujin, J.M. *Sugar Technol Rev.* **1992**, 92, 18-21.
8. Roberts, E.J. *Proc. Technical Session Cane Sugar Refining Res.,* **1980**, p 130.

Chapter 8

Aqueous Enzymatic Oil Extraction: A "Green" Bioprocess to Obtain Oil from Corn Germ and Other Oil-Rich Plant Materials

Robert A. Moreau, David B. Johnston, Leland C. Dickey, Nicholas Parris, and Kevin B. Hicks

Crop Conversion Science and Engineering Research Unit, Eastern Regional Research Center, Agricultural Research Service, U.S. Department of Agriculture, Wyndmoor, PA 19038

Several methods have been developed to obtain oil from corn germ and other oil-rich plant materials using aqueous enzymatic methods. Unlike traditional oil extraction methods, these new bioprocesses are performed without the use of presses and organic solvents. Beginning with olive oil in the pre-Biblical era, oil has been obtained from oil-rich plant materials. The large variations in cell wall ultrastructure and chemical composition (varying proportions of cellulose, hemicellulose, and pectin) of oil-rich plant materials have posed a challenge for the development of aqueous enzymatic oil extraction strategies. For most oil-rich plant materials, three types of enzymes (cellulases, proteases or pectinases) have proven to be most effective for the aqueous enzymatic oil extraction. Although the high cost of enzymes is a major hurdle to the commercialization of aqueous enzymatic oil extraction methods, recent advances in enzyme production technology may soon make the processes economically viable.

Introduction

Most plant oils are obtained by either pressing or hexane extraction. Typical oil yields from pressing range from 50-90%, depending on the type of plant material being pressed. Pressing requires much energy and the cost of energy in pressing can be a major factor in the overall cost (*1*). The oil yields with hexane extraction are typically higher, but the use of hexane has come under increasing and costly regulatory control because of safety and health concerns. Several strategies have been developed to employ aqueous, or aqueous and enzymatic methods, to extract oil from oil-rich plant materials, without the use of presses and organic solvents. The topic of aqueous oil extraction and aqueous enzymatic oil extractions were thoroughly described in two excellent reviews (*1-2*). We define aqueous oil extraction (AOE) as an extraction technology (which may employ acids, bases, buffers, organic and inorganic compounds, but no organic solvents or enzymes) to extract oil from oil-rich plant material (Table I). In contrast, we define aqueous enzymatic oil extraction (AEOE) as an extraction technology (which may employ acids, bases, buffers, organic and inorganic compounds, and enzymes, but no organic solvents) to extract oil from oil-rich plant material. Our intent in this book chapter is to provide the readers with a summary of our own recently published research in this field (*3*), and provide an update on this topic, which includes additional relevant papers that have been published in the 12-14 years since the two previous reviews *(1-2).*

Although this chapter will focus on aqueous enzymatic oil extraction (AEOE) methods, and aqueous oil extraction (AOE) methods, we acknowledge the research and potential applications which have been developed for two similar enzymatic technologies: enzyme-assisted pressing (EAP) and enzyme-assisted hexane extraction (EAHE). In a later section we describe several studies that have investigated enzyme-assisted pressing (EAP) methods to enhance the yields of olive oil. Recently Baranza *et al.* (*4*) reported that the yields of carotenoids from marigold blossoms could be increased using an enzyme-assisted hexane extraction (EAHE) method. Marigold flowers are not "oil-rich" but because the process involved enzymes and extraction, we felt it was noteworthy. Sometimes it is difficult to tell from the title of a research paper which of these aqueous and/or enzymatic technologies have been employed. For this reason, each time that a research report is described in the text or tables, we will state whether it employs an AEOE or related method (Table I).

Because of the safety, environmental, and heath issues associated with the use of hexane, the construction and operational costs of hexane extraction facilities are high. In 2001, the US Environmental Protection Agency issued

Table I. A Comparison of Conventional Extraction Technologies with Aqueous and Enzymatic Extraction Technologies

Extraction Technology (Abbrev)	*Uses Pressing*	*Uses Hexane*	*Water Added*	*Uses Enzymes*	*Uses Centrifugation*
	Conventional Processes				
Pressing/Expelling (PE)	Y	N	N[1]	N	N[1]
Hexane Extraction (HE)	N	Y	N	N	N
Prepressing/Hexane Extraction (PPHE)	Y	Y	N	N	N
Expanding/Hexane Extraction (EHE)	N	Y	N	N	N
	Experimental Processes				
Aqueous Oil Extraction (AOE)	N	N	Y	N	Y
Aqueous Enzymatic Oil Extraction (AEOE)	N	N	Y	Y	Y
Enzyme-Assisted Pressing (EAP)	Y	N	Y	Y	N
Enzyme-Assisted Hexane Extraction (EAHE)	N	Y	Y	Y	N

NOTE: [1] Used sometimes with olive oil pressing.

stricter guidelines for hexane emissions by vegetable oil extraction facilities (*5*), providing new incentives to develop alternative methods of edible oil extraction.

Historically, an aqueous process was the first method used to produce an edible plant oil in the Mediterranean region in pre-Biblical times (*6*). After harvesting, the olives were drenched in hot water and crushed with stone or wooden mortars or beam presses (Figure 1). The oil was separated from the water in a vat from which the water was removed, and the oil was then stored in jars similar to wine jars. Today, Virgin and Extra Virgin olive oils are obtained by pressing the moist olive fruit, and other grades of olive oil are obtained by solvent extraction.

Figure 1. The manufacture of olive oil via stone pressing, drawn and engraved by J. Amman in the Sixteenth Century. Unlike the pressing of dry oilseeds, water is often added before or after the pressing of olive fruits, and recently enzymes have been added to the water to enhance pressing yields (59-62). Figure is in the public domain and was provided courtesy of http://en.wikipedia.org.

Ultrastructure and Physical Properties of Various Oil-Rich Plant Materials and a Comparison of Various Enzymes to Disrupt Their Structure

In considering processes to disrupt the structure of oil-rich plant materials and release their oil contents, the contribution of the various cell wall components to the cell wall integrity needs to be considered. The main structural components of plant cell walls are cellulose, hemicellulose, pectin, lignin, and protein (*2*). Plant tissues exhibit a large range in the proportions of these major cell wall components. For example, the cell walls in soybean seeds are mainly pectin and hemicellulose (*7*), whereas in corn kernels hemicellulose is the major component, followed by cellulose, and no pectin (*8*). Most enzymatic methods that have been developed for oil extraction have employed cellulases, hemicellulases (mainly xylanases), pectinases, or proteases. However, when considering enzymatic strategies for oil extraction there is no perfect enzyme or enzyme mixture that will ensure optimal yields of oil from oil-rich material. Cellulases have been the most popular classes of enzymes used for oil extraction, but pectinases and proteases have also been effective for some oil-rich plant materials (*2*).

The oil (mainly triacylglycerols) in oil-rich plant materials is localized in organelles named lipid bodies (also sometimes called spherosomes or oleosomes) (*9-10*). Electron microscopy (Figure 2) has revealed that there is a large range in the size of lipid bodies in various plant materials (Table II). In general, lipid bodies from seeds (and milling fractions obtained from seeds) range in size from about 0.5 to 2 μm, whereas lipid bodies from oil-rich fruits are much larger (10-20 μm).

Like other organelles, the lipid bodies are surrounded by a biological membrane, but the structure and properties of this membrane have two unusual features. First, instead of being comprised of a typical "phospholipid bilayer", USDA scientists first provided evidence in the 1960s that the lipid body membrane is actually a phospholipids monolayer (*11-12*). In addition, Huang and co-workers (*13-17*) discovered that the membrane of the lipid body of most seeds contain a unique group of structural proteins which they called "oleosins." It has been postulated that a major function of the oleosins is to protect the integrity of the oil body and prevent its contents from coalescing during dehydration (which occurs during seed formation) and rehydration (which occurs during seed germination) (*18*). The size of most oilseed oleosins is 15-26 KDa. The study of the amino acid sequences of several oleosins has revealed that each has a hydrophobic region, which is in contact with the phospholipids in the monolayer of the oil body membrane and a larger hydrophilic region, which protrudes from the outer surface of the oil body. Tze and Huang (*13*) reported that the phospholipids in intact oil bodies are resistant to hydrolysis by phospholipase C and phospholipase A2, but the oleosin protein is susceptible to hydrolysis by trypsin. Based on the results of these enzymatic hydrolyses, it was postulated that the large hydrophilic region in oleosin formed a mushroom-shaped covering of the surface of the oil body. making it impossible (due to steric hindrance) for phospholipases to reach the phospholipids in the oil body membrane (Figure 3) (*13*).

Unlike the oil bodies in most seeds, those in oil-rich fruits (*e.g.*, olive, avocado) have little or no oleosins, perhaps because these fruits do not undergo dehydration and rehydration (*19*). In addition to oleosins serving to maintain the physical integrity of the oil bodies during dehydration and rehydration, there is also evidence that they may provide binding sites for lipases during seed germination (*10*). Because oleosins contribute to the barrier properties of oil

Table II. A Comparison of the Ultrastructure and Physical Properties of Oil-Rich Plant Materials

Type of Oil Rich Plant Material	*Example*	*Monocot or Dicot*	*Ave Diameter of Oil Body (μm)*	*Oleosins in Oil Body*	*Ref*
Oilseed Crops					
	Soybean	D	NR	Y	*(17)*
	Rapeseed/canola	D	0.65	Y	*(13)*
	Sunflower	D	1.5	Y	*(15)*
	Peanut	D	1.95	Y	*(14)*
	Cottonseed	D	0.97	Y	*(14)*
	Palmkernel	M	NR	Y	*(17)*
	Castor	D	NR	Y	*(17)*
Grain Processing Fraction					
	Corn Germ	M	1.45	Y	*(13)*
	Rice Bran	M	NR	Y	*(16)*
	Wheat Germ	M	NR	Y	*(17)*
Fruit (mesocarp)					
	Olive	D	10-20	N	*(19)*
	Avocado	D	10-20	N	*(19)*
	Palm	M	NR	N	NR

NOTE: Abbreviation, NR, not reported.

bodies in seeds, strategies need to be developed to break (breech) this physical barrier and release seed oil. This may be accomplished by physical (pressing), thermal (cooking), chemical (SO_2), or enzymatic (proteases) strategies.

Light microscopy and scanning electron microscopy have been used to study the "microstructural features of enzymatically treated oilseeds" (*20*). Although, some of the gross changes caused by enzymes were documented, the effect of enzymes on the oil bodies or oleosins were not considered *(20)*.

Conventional Oil Extraction Processes

The oldest technique for the removal of oil from oil-rich plant material involves pressing (Table I). In ancient times olives were pressed with a wooden

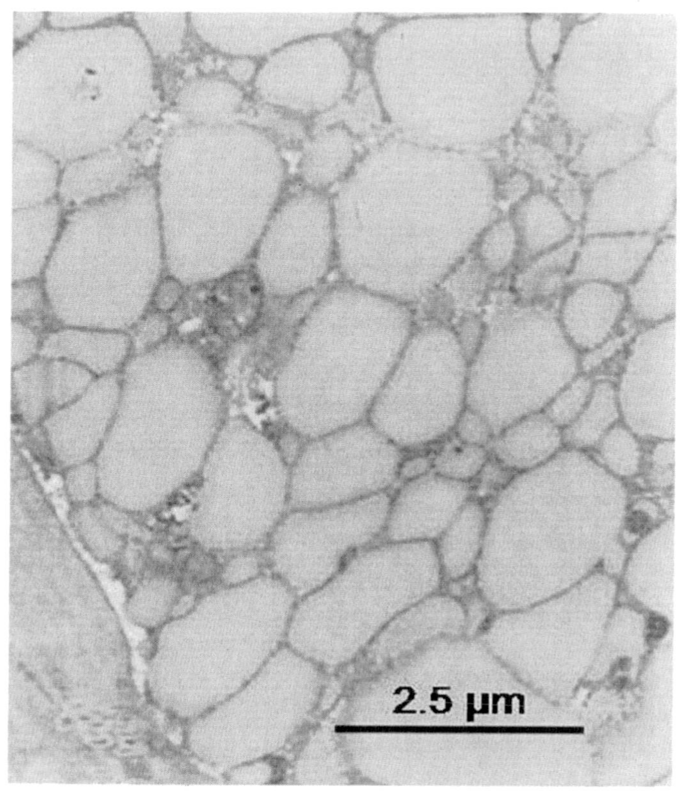

Figure 2. Transmission electron microgram of a cross section of dry milled corn germ showing oil bodies with an average diameter of ~1 µm. Micrograph was generously provided by Dr. Peter Cooke.

pole or stone press (Figure 1). Pressing is a simple technique, but is unable to achieve high yields of oil. "Cold pressing" is conducted at low temperatures to preserve the flavor components. Today, cold pressing is usually only used to produce two high-value oils: olive and sesame oils. A common method for efficiently producing commodity oils from oilseeds involves mechanical expelling ("prepressing") and hexane extraction (Table I) (*21*). Extrusion (with a specialized extruder named an expander) has also been employed as a means of germ preparation for solvent extraction (see EHA in Table 1), producing a crude corn oil of high quality and high yield (*21*). Hexane is removed from the oil-rich miscella by evaporation, heat, and partial vacuum, and the hexane is condensed recycled. Hexane is also evaporated and recycled from the germ cake. Corn germ meal contains 23-25% protein, is usually sold as an animal feed ingredient, and is often added to increase the protein content of corn gluten feed.

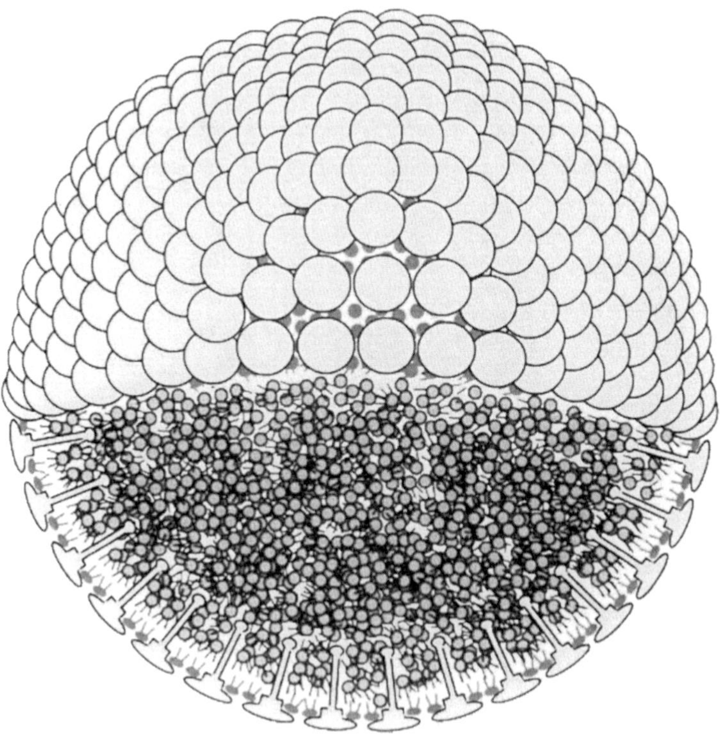

Figure 3. A model of the maize oil body. Note how the oleosins cover the outer surface of the membrane, shielding the phospholipids. Drawing courtesy of Dr. Anthony Huang.

Because of the safety, environmental, and heath issues associated with the use of hexane, ethanol has also been proposed as an organic solvent for the extraction of corn oil from corn germ or from whole kernels. Some advantages of ethanol over hexane are its higher flash point, its "food-grade" qualification, and that it can be readily produced from corn via fermentation. Some disadvantages include its higher boiling point (requiring more energy to remove solvent from meal and miscella), and its increased polarity which means that it extracts more polar extractants (*e.g.*, phospholipids) that may need to be removed during refining. Hojilla-Evangelista *et al.* (*22-23*) developed a process named "the sequential extraction process (SEP)," which involves the ethanol extraction of whole flaked corn kernels to first extract the corn oil, and then additional extraction with ethanol to extract some of the proteins (zein) and starch. The economics of the SEP process have been rigorously evaluated, and some recent process modifications have been proposed to improve the efficiency and lower the cost (*22-24*). Recently, a method was developed to extract corn oil and zein from ground corn with ethanol, using a proprietary process which employs membrane filters (*25-26*). Another potential problem with using ethanol to extract oil is that oil has limited solubility in ethanol, compared to hexane.

Supercritical CO_2 extraction methods were evaluated for a number of oilseeds in the 1980s. Corn germ extraction by supercritical CO_2 was evaluated by ARS researchers in Peoria, IL. Methods for corn germ extraction using 100% supercritical CO_2 were developed (*27*) and patented (*28*). Others have demonstrated that addition of ethanol modifier (0-10%) to supercritical CO_2 can decrease the extraction time and improve the functionality of germ proteins (*29*). Although there are no technical barriers, extraction of corn germ with supercritical CO_2 is much more costly than conventional extraction methods.

AEOE Methods to Extract Corn Oil from Corn Germ

In 1994, a Yugoslavian group reported an AEOE method that resulted in an 80% yield of corn oil from wet milled corn germ (Table III) using commercial enzyme preparations (*8,30*). In their first paper, the group used Pectinex Ultra SP-L, a pectinolytic enzyme preparation from *Aspergillis niger* (*30*), and in their second paper the group obtained higher yields using Cellulclast^{TM}, a cellulase enzyme preparation from *Trichoderma reseei* (*8*). Both processes started with corn germ from wet milling, and the corn germ was used while still wet, and not oven dried. Part of the process involved "hydrothermal" treatment (the equivalent of an autoclave treatment of the corn germ) prior to the addition of enzyme.

Table III. A Comparison of the Three Published AEOE Methods for the Extraction of Corn Oil from Corn Germ. Wet Milled Corn Germ was Used in All Studies

Germ Dried in Oven? (% moisture)	*Hydrothermal Processing Step (112 °C)*	*Enzyme used*	*Oil Yield*	*Oil Quality*	*Ref*
N (53%)	Y	Pectinex Ultra SP-L	NR	Excellent	*30*
N (53%)	Y	Celluclast	80%	NR	*8*
Y (3%)	N	Celluclast or Multifect GC or GC 220	80-90%	Excellent	*3*

NOTE: Abbreviation: NR, not reported.

Our laboratory recently developed an AEOE method (Table III) to extract corn oil from wet milled corn germ (*3*). The basic steps in the method involved "churning" the corn germ with various enzymes and buffer for 4 h at 50 °C, and an additional 16 h at 65 °C, followed by centrifugation and removal of the oil layer from the surface. No hexane or other organic solvents are used in this process. Using oven dried corn germ samples (6 g) from a commercial corn wet mill, corn oil yields of ~80% (relative to hexane extraction) were achieved using three different commercial cellulases (Figure 4). Using this method in the absence of enzyme resulted in corn oil yields of ~37% (*3*). A four-fold scale up of the method (to 24 g of germ) resulted in oil yields of ~90%. Most of the remaining 10-20% of oil is probably in a white emulsion layer located directly below the floating oil layer after centrifugation (Figure 4).

Three commercial cellulases (Multifect GC and GC 220 from Genencor, and Celluclast from Novozyme) were evaluated using our protocol and all three resulted in oil yields of ~80% (*3*). An additional cellulase (Sigma C1794) and two xylanases (Multifect Xylanase from Genencor and Sigma X 2753) resulted in yields in the range of 54 to 66%. Six other enzymes (including three cellulases, two proteases and one pectinase) resulted in yields of ~30-44%, whereas an oil yield of ~27% was achieved using this protocol with no enzyme. Six of the seven most effective enzymes were from the fungus *Trichoderma*. Although these enzymes are marketed as cellulases and xylanases, all of them are actually mixtures of enzymes and contain other enzyme activities that may contribute to their ability to extract corn oil. The levels of seven different hydrolytic enzymes (protease, cellulase, β-glucanase, xylanase, hemicellulase, amylase, and native starch amylase) were recently compared in a commercial cellulase preparation and a commercial xylanase preparation from *Trichoderma reesei* (*31*).

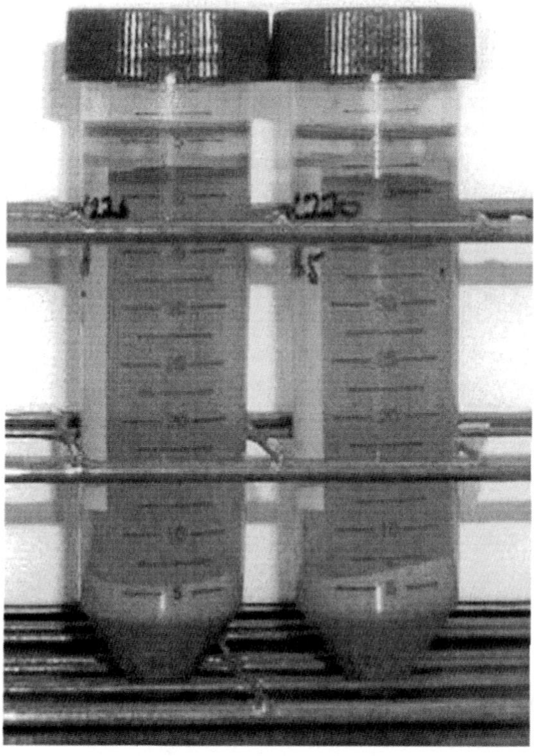

Figure 4. Aqueous Enzymatic Oil Extraction of 6 grams of Wet Milled Corn Germ using the method reported for wet milled corn germ (3). Note the clear oil layer on top (~2.4 g) and white interface emulsion layer below it.

Karlovic *et al.* (*8*) and Singh *et al.* (*32*) both noted that unlike most oilseeds, arabinoxylans are the most abundant carbohydrate polymer in corn germ. Because of this, it is reasonable that enzyme preparations that combine xylanase and cellulase activities may be the most effective (Table II). Jacks *et al.* (*11-12*), provided evidence that lipid bodies, the triacylglycerol-containing organelles in seeds, are surrounded by a "half unit" membrane that is comprised of a phospholipid monolayer and a structural protein called "oleosin." Therefore, it is reasonable to hypothesize that proteases and/or phospholipases may be useful enzymes for aqueous enzymatic oil extraction. Indeed, Hanmoungjai *et al.* (*33*) reported that Alcalase (a protease) was useful for the aqueous enzymatic extraction of oil from rice bran. However, the two proteases tested in our study (*3*) had almost no effect on oil yields. To our knowledge, no one has evaluated phospholipases for their efficacy at enzymatic oil extraction. However, because phospholipases could potentially degrade all cellular biomembranes, and some phospholipases could release lysopholipids (which are known to act as surfactants), their use may be problematic. Also, Tze and Huang (*13*) reported that phospholipase C and phospholipase A2 had no effect on the membrane phospholipids in isolated corn lipid bodies. They postulated that the large hydrophobic portion of the oleosin protein formed a "shield" over the outer surface of the oil body membrane and prevented the phospholipases from gaining access to the membrane phospholipids (Figure 3), thus providing a steric effect (*13*).

When the oil quality of hexane-extracted versus our aqueous enzymatic-extracted corn oils were compared, the two compositions were very similar (*3*). Bocevska *et al.* (*30*) also reported that their AEOE corn oil was of high quality. The very low levels of free fatty acids generated during AEOE indicate that lipolytic activity was minimal in our oven dried wet milled corn germ (*3*), and in the hydrothermal treated wet milled corn germ (*8,30*). It is possible that if this aqueous enzymatic method was used with wet corn germ instead of oven-dried corn germ, then the hydrothermal treatment may then be necessary. The levels of phytosterols (free and esterified) were slightly lower in the aqueous enzyme extracted oil than in hexane extracted oil.

Our new aqueous enzymatic extraction process results in oil yields of greater than 90%. This yield is higher than the 80% yield of corn oil from corn germ previously reported by Karlovic *et al.* (*8*). Care needs to be taken in directly comparing our yields with those previously reported (*8*) because in the earlier report, wet (undried) corn germ was used as a feedstock for oil extraction, but in our present method factory dried germ was used. Also, the aqueous enzymatic extraction process used by Karlovic *et al.* (*8*) included an essential "hydrothermal pretreatment" step, whereas our method resulted in high yields without a hydrothermal pretreatment step. However, as mentioned previously, if we had used wet (fresh, not heat-dried) corn germ, it is possible that the hydrothermal pretreatment may have been necessary. It should be noted that no precautions were taken to limit the growth of microbes during our aqueous enzymatic process. We recognize that the development of a successful aqueous enzymatic oil extraction process will probably require the implementation of strategies to limit microbial growth. Finally, since several cellulase preparations appear to result in high oil yields, we anticipate that some of the new generation of celluloytic enzymes that are being developed for biomass hydrolysis and fermentation, may result in even higher oil yields and may be less costly to produce.

AEOE Methods for Oilseeds and Grain Milling Fractions

Several AEOE methods for oilseeds and grain milling fractions were reported in the 1996 AEOE review (*2*). In Table IV we summarize the AEOE methods for oilseeds that have been reported since the 1996 review (*7*, *34-56*). For comparative purposes (and because it is often difficult to distinguish from the title of a paper the extraction method being studied) we have also included a few recent EAHE, EAP and AOE methods. Although coconut is sometimes considered a "fruit" (*1*), from a botanical standpoint the oil bodies in the coconut are contained within the large seed so, for the sake of this discussion, we have grouped it with oilseeds.

A consideration of the recent oilseed AEOE results indicates that both cellulases and proteases have been successfully employed. It should be noted that although commercial enzymes are marketed with the names protease and cellulase, almost all commercial enzymes are mixtures of multiple enzymes, usually from microbial sources.

AEOE Methods for Oil-Rich Fruits

This section focuses on avocado, olive, and palm, the three major crops where oil is obtained from the "fruit." The botanical definition of "fruit" is defined as the fleshy layers (mesocarp) that surround the seed. It has been reported that the oil bodies of avocado and olive have two features that distinguish them from the oil bodies of oilseeds (*19*). First, they are much larger (diameters of 10-20 μm) than seed oil bodies (which have diameters of 0.5 to 2 μm). Also, unlike the membranes of seed oil bodies, the membranes of the oil bodies of avocado and olive do NOT contain oleosins (*19*). Similarly, it is likely that the membranes of the oil bodies of palm and other oil-rich fruits are also devoid of oleosins (personal communication, Dr. Anthony Huang). The absence of oleosins as a physical barrier in the membranes of the fruit lipid bodies may mean that the oil is easier to extract from fruit using AEOE methods than it is to extract oil from seeds. Oil bodies have been isolated from palm mesocarp, but their diameters and the occurrence of oleosins in the oil body membranes was not reported (*57*).

We are only aware of one AEOE method for avocados (*58*), where oil yields of 65-68% were reported with amylase or a mixture of protease + cellulase.

Although we are not aware of any AEOE methods for olives, several studies have investigated enzyme-assisted pressing (EAP) methods to enhance the yields of olive oil. The absence of AEOE methods for olive may be attributed to the high value of Virgin and Extra Virgin olive oils, and the desire to develop enzymatic processes to improve the yields of these valuable products. Olive

Table IV A. A Comparison of Some Recently Published AEOE Methods for Oilseeds and Oil-Rich Grain Milling Fractions

Oilseed or other Plant Material	*Method Type*	*Enzyme Used*	*pH*	*Hours*	*Oil Yields (wt %)*	*Ref*
Soybean	Extrusion + AEOE	Celluclast 1.5L [c] + Alcalase 2.4L [pr]	6-7	6	88	*(34)*
	AEOE	Noncommercial from *Aspergillis*	NR	8	79	*(35)*
	AOE	none	8-12	10	90	*(36)*
	EAP + EAHE	Driselase [pr] and other enzymes	3-5	16	88	*(37)*
	AEOE	Alcalase 2.4 L [pr]	8-9	1	59	*(7)*
Peanut	AOE	none	4	16	80	*(38)*
	AEOE	Noncommercial from *Aspergillis*	NR	14	53	*(39)*
	AEOE	Protizyme [pr]	4.0	18	86-92	*(40)*
Rapeseed/ canola	AEOE	Noncommercial from *Aspergillis*	NR	8	41	*(41)*
	AEOE	Noncommercial from *Aspergillis*	NR	8	44	*(42)*
	AEOE	SP 331	4.5	4	80-85	*(43)*
Sunflower	AEOE	Celluclast 1.5L [c] + Pectinex ULTRA SP [pe]	6.7	10	90	*(44)*
	AEOE	Noncommercial from *Aspergillis*	NR	14	57	*(39)*
Coconut	AEOE	Celluclast[c] + Viscozyme[pe] + Alcalase[pr] + Termamyl[a]	7	1	74	*(45)*
	AEOE	Sumizyme LP [pr] + Sumizyme C [c]	6	12	65	*(46)*
Plum Kernel	AEOE	Viscozyme 120 L [pe]	4.5	3	70	*(47-48)*

NOTE: Abbreviations: [pr], protease; [c], cellulase; [a], amylase, [a]; [pe], pectinase, [e], oil in emulsion; NR, not reported

Table IV B (Continued). Comparison of Some Recently Published AEOE Methods for Oilseeds and Oil-Rich Grain Milling Fractions

Oilseed or other Plant Material	*Method Type*	*Enzyme Used*	*pH*	*Hours*	*Oil Yields (wt %)*	*Ref*
Chilean Hazelnut	AEOE	Olivex + Celluclast [c]	6.2	12	60+30[e]	*(49)*
Shea	AEOE	Sumizyme AP [pr] + Sumizyme C [c]	5.5	6	69	*(50)*
Cocoa	AEOE	Sumizyme LP [pr] + Sumizyme C [c]	NR	6	73	*(51)*
Moreinga oleifera	AEOE	Neutrase 0.8L [pr]	6.8	24	72	*(52)*
Jatropha curcas	AEOE	Protizyme [pr]	9	6	74	*(53)*
Rice Bran	EAHE	Pectinex ULTRA SPL[pe]	4.5	4	85	*(54)*
	AEOE	Celluclast R [c]	4.5	4	33	
	AOE	none	12	1	85	*(55)*
	AEOE	Alcalase 0.6L [pr]	9	3	79	*(33)*
	AEOE	Amylase + Cellulase + Protizyme[pr]	8	20	77	*(56)*

NOTE: Abbreviations: [pr], protease; [c], cellulase; [a], amylase, [a]; [pe], pectinase, [e], oil in emulsion , NR, not reported

fruits are usually pressed when they still contain a high level of moisture, and sometimes water is added before or after pressing, so the addition of enzymes to this aqueous system is a reasonable modification. Ranalli and Serracicco (59) demonstrated increased extraction yields using a cellulase mixture from *Aspergillus aculeatus*. Vierhuis *et al.* (*60*) investigated the use of a pectinolytic enzyme mixture from *Aspergillus niger*. They demonstrated that it both increased the yields of virgin olive oil and the levels of two secoiridoid derivatives. Rannalli *et al.* (*61*) reported that an enzyme mixture (containing cellulase, hemicellulases and pectinases) produced better quality oils and increased yields. Another recent paper reported enhanced quality of virgin olive oil by the use of "Bioliva," a novel enzyme extract obtained from plants (*62*).

Although palm oil is the #2 oil crop in the world (*63*), slightly behind soybeans (23 versus 27 million metric tonnes, respectively, in 2001) no published information could be found about oil bodies, oleosins, or attempts at developing AEOE methods for oil palm.

Optimizing the Value of Coproducts and Byproducts of AEOE

In considering the economic feasibility of all AEOE methods, the potential utilization of all non-oil components needs to be explored and optimized. Traditionally, the byproducts of oil pressing and hexane extraction have been sold as animal feed. For most oilseeds and oil-rich fruits, higher profits are obtained from the sale of oil than from the sale of its coproducts. In the case of soybean, which contains about 20% oil, more profits are usually obtained from the sale of the protein-rich meal coproduct than from the sale of the oil (*2*). In our corn germ AEOE research program, we are investigating several high-value applications for coproducts. These include food-grade gum, and industrial applications for corn germ arabinoxylans and food and feed applications for corn germ proteins. In addition, we are investigating peptides that could have high-value applications as antimicrobial compounds, prebiotics, and anti-hypertension agents (for lowering blood pressure).

The Future of AEOE

Because of the safety and environmental issues associated with the use of hexane, the construction and operational costs of hexane extraction facilities are high. In 2001, the US Environmental Protection Agency issued stricter guidelines for hexane emissions by vegetable oil extraction facilities (*5*), providing new incentives to develop alternative methods of edible oil extraction.

Our current corn germ AEOE project was undertaken to evaluate the possibility of obtaining corn oil from corn germ by applying some of the previously-published methods for the aqueous and aqueous enzymatic extraction of oil from oilseeds. In 1994, Karlovic *et al.* (*8*) estimated that corn oil produced by their corn germ AEOE method would cost 2.5 times more than corn oil produced by conventional extraction. They also estimated that the cost of enzymes would constitute about 90% of the cost of their AEOE method. Olsen (*44*) estimated that the cost to produce rapeseed oil via AEOE was 5-6 times higher than to produce it by conventional hexane extraction. Most other cost estimates of AEOE methods for other oil-rich plant materials have reached similar conclusions. However, since much effort is being devoted to developing less-costly enzymes to hydrolyze and ferment cellulosic biomass to ethanol (*64*), there will soon be an opportunity to evaluate some of these newly developed enzymes for the extraction of corn germ and other oil-rich materials. Also, the current high demand for corn-derived fuel ethanol is creating a surplus of corn germ, and the need for new extraction technologies may create a demand for the development of new and cheaper enzymes.

Conclusions

Several reports have demonstrated the technical feasibility of AEOE methods. The major hurdle for the commercialization of these processes is enzyme cost. With the rapid development of many new and less expensive cell-wall degrading enzymes, we are hopeful that research will continue to gradually reduce the cost of oil produced by AEOE methods. In addition to the environmental, safety, and health benefits of AEOE methods (mainly due to the elimination of the need for hexane), another advantage of AEOE methods is that phospholipids would be removed during the extraction process, thus eliminating the need to need for a degumming step during the oil refining process.

References

1. Dominguez, H.; Nunez, M. J.; Lema, J. M. *Food Chem.* **1994**, *49,* 271-286.
2. Rosenthal, A.; Pyle, D. L.; Niranjan, K. *Enzyme Microb. Technol.* **1996**, *19,* 402-420.
3. Moreau, R. A.; Johnston, D. B.; Powell, M. J.; Hicks, K. B. *J. Am. Oil. Chem. Soc.* **2004**, *81*, 1071-1075.
4. Barzana E.; Rubio, D.; Santamaria, R. I.; Garcia-Correa, O.; Garcia, F.; Ridaura Sanz, V. E.; Lopez-Munguia, A. *J. Agric. Food Chem.* **2002**, *50,* 4491-4496.
5. Environmental Protection Agency; 40 CFR Part 63; *National Emissions Standards for Hazardous Air Pollutants: Solvent extraction for vegetable oil production; Final Rule. Federal Register* **2001**, *66,* 19005-19026.
6. Firestone, D. In *Bailey's Industrial Oil and Fat Products;* Shahidi Ed.; Sixth Edition; Wiley-Interscience: Hoboken, **2005**; *Vol. 2*, p 149-172.
7. Rosenthal, A.; Pyle, D. L.; Niranjan, K. *Enzyme Microb. Technol.* **2001**, *28*, 499-509.
8. Karlovic, D. J.; Bocevska, M.; Jakolevic, J.; Turkulov, J. *Acta Alimentaria* **1994**, *23,* 389-400.
9. Huang, A. H. C. *Annu. Rev. Plant Physiol.Plant Mol. Biol.* **1992**, *43*, 177-200.
10. Huang, A. H. C. *Plant Physiol.* **1996**, *110,* 1055-1061.
11. Jacks, T. J., Yatsu, L. Y.; Altschul, A. M. *Plant Physiol.* **1967**, *42*, 585-597.
12. Jacks, T. J., Hensarling, T. P.; Neucere, J. N.; Yatsu, L. Y.; Barker, R. H. *J. Am. Oil Chem. Soc.* **1990**, *67*, 353-361.
13. Tzen, T. C.; Huang, A. H. C. *J. Cell. Biol.* **1992**, *117*, 327-335.
14. Tzen, T. C.; Cao, Y. –Z.; Laurent, P.; Ratnayake, C.; Huang, A. H. C. *Plant Physiol.* **1993**, *101*, 267-276.
15. Beisson, F.; Ferte, N.; Noat, G. *Biochemical J.* **1996**, *317*, 955-956.

16. Chuang, R. L. C.; Chen, J. C. F.; Chu, J.; Tzen, J. T. C. *J. Biochem.* **1996**, *120*, 74-81.
17. Tzen, J. T.; Lai, Y.-K., Chan, K.-L, Huang, A. H. C. *Plant Physiol.* **1990**, *94*, 1282-1289.
18. Leprince, O., van Aelst, A. C., Pritchard, H. W.; Murphy, D. J. *Planta* **1998**, *204*, 109-119.
19 Ross, J. H.; Sanchez, E. J.; Millan, F.; Murphy, D. J. *Plant Sci.* **1993**, *93*, 203-210.
20 Sineiro, J.; Dominguez, H.; Nùnez, M. J.; Lema, J. M. *J. Sci. Food. Agric.* **1998**, *78*, 491-497.
21. Moreau, R. A. In *Bailey's Industrial Oil and Fat Products*; Shahidi, F. Ed.; Sixth Edition; Wiley-Interscience: Hoboken, **2005**, *Vol. 2*, p 149-172.
22. Hojilla-Evangelista, M. P.; Johnson, L. A.; Myers, D. J. *Cereal Chem.* **1992**, *69*, 643-647.
23. Hojilla-Evangelista, M. P.; Johnson, L. A. *J. Am. Oil Chem. Soc.* **2002**, *79*, 815-823.
24. Feng, F.; Myers, D. J.; Hojilla-Evangelista, M. P.; Miller, K. A.; Johnson, L. A.; Singh, S. K. *J. Am. Oil. Chem. Soc.* **2002**, *79*, 707-709.
25. Kwiatkowski, R; Cheryan, M. *J. Am. Oil Chem. Soc.* **2002**, *79*, 825-830.
26. Cheryan, M. *U.S. Patent 6,433,146*, Corn oil and protein extraction method (University of Illinois), August 13, **2002**.
27. Christianson, D. D.; Friedrich, J. P.; List, G. R.; Warner, K.; Bagley, E. B.; Stringfellow, A. C.; Inglett, G. E. *J. Food Sci.* **1984**, *49*, 229-233.
28. Christianson, D. D.; Friedrich, J. P. *U.S. Patent 4,495,207* (U.S. Department of Agriculture) Production of food-grade corn germ product by supercritical fluid extraction, January 22, **1985**.
29. Rónyai, E.; Simándi, B.; Tömösközi, S.; Deák, A.; Vigh, L.; Weinbrenner, Z. *J. Supercrit. Fluids* **1998**, *14*, 75-81.
30. Bocevska, M., Karlovic, D.; Turkulov, J.; Pericin, D. *J. Amer. Oil Chem. Soc.* **1993**, *70,* 1273-1277.
31. Johnston, D. B.; Singh, V. *Cereal Chem.* **2001**, *78,* 405-411.
32 Singh, V.; Doner, L. W.; Johnston, D. B.; Hicks, K. B.; Eckhoff, S. R. *Cereal Chem.* **2000**, *77*, 560-561.
33 Hanmoungjai, P.; Pyle, D. L.; Niranjan, K. *J. Amer. Oil Chem. Soc.* **2001**, *78,* 817-821.
34. Fretias, S. P.; Hartman, L.; Couri, S.; Jablonka, F. H.; de Carvalho, C. W. P. *Fett/Lipid* **1997**, *99,* 333-337.
35. Kashyap, M. C.; Agrawal, Y. C.; Sarkar, B. C.; Singh, B. P. N. *J. Food Sci.* **1997**, *34*, 386-390.
36 Rosenthal, A.; Pyle, D. L.; Niranjan, K. *Trans IchemE* **1998**, *76,* 224-230.
37. Bargale, P. C. *J. Food Proc. Eng.* **2000**, *23,* 321-327.
38. Shi, L.; Lu, J.; Jones, G.; Loretan, P. A.; Hill, W. A. *Life Support and Bioscience* **1998**, *5,* 225-229.

39. Singh, R. K.; Sarker, B. C.; Kumbhar, B. K. *J. Food Sci. Technol-Mysore* **1999**, *36*, 511-514.
40. Sharma, A., Khare, S. K.; Gupta, M. N. *J. Amer. Oil Chem. Soc.* **2002**, *79*(3), 215-218.
41. Sarker, B. C., Singh, B. P. N., Agrawal, Y. C.; Gupta, D. K. *J. Food Sci.* **1998**, *35*, 183-186.
42. Srivastava, B.; Agrawal, Y. C.; Sarker, B. C.; Kushwaha, Y. P. S.; Singh, B. P. N. *J. Food Sci. Technol.* **2004**, *41*, 88-91.
43. Olsen, Hans Sejr, Aqueous Enzymatic Extraction of Oil from Rapeseeds, a case study by Novo Nordisk A/S, http://www.p2pays.org/ref/10/09365.htm, **2006**.
44. Dominguez, H.; Sineiro, J.; Nunez, M. J.; Lema, J. M. *Food Res. Int.* **1996**, *28,* 537-545.
45. Man, C.; Suhardiyono, Y. B.; Asbi, A. B.; Azudin, M. N.; Wei, L. S. *J. Am. Oil Chem. Soc.* **1996**, *73,* 683-686.
46. Tano-Debrah, K.; Ohta, Y. *J. Sci. Food Agric.* **1997**, *74,* 497-502.
47. Picuric-Jovanovic, K.; Vrbaski, Z.; Milovanovic, M. *Fett-Lipid* **1997**, *99(12),* 433-435.
48. Picuric-Jovanovic, K.; Vrbaski, Z.; Milovanovic, M. *Fett-Lipid* **1999**, *101*, 109-112.
49. Santamaria, R. I.; Soto, C.; Zuniga, M. E.; Chany, R.; Lopez-Munguia, A. *J. Am. Oil Chem. Soc.* **2003**, *80,* 33-36.
50. Tano-Debrah, K.; Yoshimura, Y.; Ohta, Y. *J. Am. Oil Chem. Soc.* **1996**, *73*, 449-453.
51. Tano-Debrah, K.; Ohta Y. *J. Am. Oil Chem. Soc.* **1995**, *72*, 1409-1411.
52. Abdulkarim, S. M.; Long, K.; Lai, O. M.; Muhammad, S. K. S.; H. M. Ghazali. *Food Chem.* **2005**, *93*, 253-263.
53. Shah, S.; Sharma, A.; Gupta, M. N. *Bioresource Technol.* **2005**, *96*, 121-123.
54. Sengupta, R.; Bhattacharyya, D. K. *J. Am. Oil Chem. Soc.* **1996**, *73*, 687-692.
55. Hanmoungjai, P., Pyle, L.; Niranjan, K. *J. Chem. Technol. Biotechnol.* **2000**, *75,* 348-352.
56. Sharma, A., Khare, S. K.; Gupta, M. N. *J. Am. Oil Chem. Soc.* **2001**, *78,* 949-951.
57 Oo, K.-C.; Chew, Y.-H. *Plant Cell Physiol.* **1992**, *33*, 189-195.
58. Buenrostro, M.; Lopez-Munguia, C. *Biotechnology Letters* **1986**, *8*, 505-506.
59. Ranalli A.; Serraiocco A. *Grasas y Aceites* **1996**, *47*, 227-236.
60. Vierhuis, E.; Servili M.; Baldioli M. *J. Agric. Food Chem.* **2001**, *49*, 1218-1223.
61. Ranalli A.; Ferrante M. L.; De Mattia, G. *J. Agric. Food Chem.* **1999**, *47*, 417-424.

62. Ranalli, A.; Pollastri, L.; Contento, S.; Lucera, L.; Del Re, P. *Eur. Food Res. Technol.* **2003**, *216,* 109-115.
63. Gunstone, F. D. In *Properties and Uses*, Blackwell Publishing: Oxford, **2002**; p 333.
64. McFarland, K. C.; Ding, H.; Teter, S.; Vlasenko, E.; Xu, F.; Cherry, J. In *Industrial Application of Enzymes on Carbohydrate Based Materials*; Eggleston, G.; Vercellotti, J. R.; Eds.; ACS Symposium Series: Washington D. C., **2007**, in press.

Advances in the Applications of Enzymes in the Textile Industry

Chapter 9

Application of a Thermostable Pectate Lyase in the Bioscouring of Cotton Fabrics at Laboratory and Pilot Scales

Ali R. Esteghlalian, Martin M. Kazaoka, David F. Walsh, Ryan T. Mccann, Arne I. Solbak, Janne Kerovuo, And Geoffrey P. Hazlewood

Diversa Corporation, 4955 Directors Place, San Diego, Ca 92121

The highly effective use of a commercial, thermostable pectate lyase product in removing pectin and other hydrophobic materials from cotton fabrics is demonstrated. The scouring efficiency was evaluated at the laboratory scale as well as in a pilot scale jet type unit. This enzyme product was able to effectively degrade the pectin, enhance the hydrophilicity of Interlock cotton fabric, and produce uniform and acceptable dyeing results.

Introduction

Enzymes have a wide array of applications in the textile industry, which include desizing, depilling, and bioscouring of cotton and blend fabrics, biostoning of denim fabrics and garments, and shrink proofing of wool fibers. The use of enzymes has helped textile mills to reduce their negative environmental impact by reducing chemical and biological load in the mill effluents. Enzyme applications have also enabled the textile industry to improve its economic returns, by reducing energy and water consumption in the mill and by producing more desirable products in terms of quality and appearance. For instance, enzymatic scouring (bioscouring) is an environmentally friendly alternative to caustic scouring, that can help reduce the amount of water used for the post-caustic rinses and eliminate the harsh caustic treatment that can damage fiber strength. This book chapter provides an overview of the use of various enzymes in the textile industry, and presents the results of cotton bioscouring with a commercial, thermotolerant pectate lyase at the laboratory and pilot scales.

Enzymes in Wet Processing of Cotton Fabrics

Several categories of hydrolytic enzymes, namely amylases, pectinases and pectate lyases, are used for the "preparation" of cotton yarns and fabrics. The purpose of the preparation process is to prepare the yarn or fabric for the dyeing operation. This 3-stage process of desizing, scouring, and bleaching, traditionally uses a combination of chemicals and harsh process conditions to remove the natural or man-made non-cellulosic material, such as starch, pectin, motes, natural and process-derived oils and waxes, from the fiber surface. The dye molecules can then closely and evenly associate with the cellulosic component of the well prepared fiber and produce evenly dyed yarns and fabrics (*1*).

Amylases for Desizing

Amylases were the first enzymes used by the textile industry and are widely used today in desizing operations to remove the starch from sized cotton knit fabrics. Cotton fibers are sized using film forming materials, such as starch, carboxymethyl cellulose, polyvinyl alcohol (PVA) or triglyceride, to improve the weaving process by increasing yarn strength and abrasion resistance, and to maintain yarn flexibility and elongation. In addition to the main sizing agent, a sizing bath also contains waxes or synthetic lubricants, binders, softeners, wetting agents, and other additives (*2*). In preparation for the dyeing process, sizing materials have to be removed to enhance fabric absorbency and dye uptake (*3-4*). Chemical desizing is undertaken with a mixture of soda ash and surfactants at high pH, whereas amylases can hydrolyze and remove starch at near neutral pH levels.

A variety of thermal and alkaline tolerant amylases active over a wide range of pH (5-8) and temperature (50-100 °C) are available for batch, semi-continuous, and continuous desizing (*4*). A desizing process typically consists of pre-washing, enzyme impregnation and starch hydrolysis, and after-wash stages. During the pre-washing stage, fibers are cleaned by removing waxes and water-soluble, non-starch matter to enhance enzyme-starch interaction. The starch is thoroughly hydrated and hydrolyzed in the subsequent impregnation and hydrolysis stages, where the enzyme hydrolyzes the starch into oligosaccharides that are released into the desizing bath. A final hot rinse (100 °C) combined with agitation is used to release the hydrolyzed starch, and is followed by multiple rinses and pH adjustment to neutrality. The progress of starch hydrolysis during the desizing process can be monitored by exposing the washed fabric to a mild stock of iodine solution, monitoring the color produced, and comparing the color against the standard Violet Scale (*3,5*).

The desizing bath temperature, pH, and presence of substrate and calcium ions all influence amylase stability (*4*). The substrate (starch) protects the enzyme from deactivation, especially at higher temperatures and extreme pH levels. Calcium ions are required for the proper folding and activation of most amylases, and enzymatic desizing baths typically contain NaCl (16-18%), calcium ions (~0.5 g/L $CaCl_2$), and surfactants to facilitate the hydration and hydrolysis of starch. Calcium can also be added as a formulation ingredient in commercial amylase products (*4*).

Cellulases for Biostoning and Biopolishing

Cellulases are widely used in the garment industry to create desirable "worn-out", abraded, or stonewashed effects on denim garments. Before their introduction into the garment industry, these fashionable effects were achieved by washing the denim garments with pumice stones or treatment with harsh chemicals such as potassium permanganate and sodium hypochlorite (*2,6*). The use of pumice and abrading chemicals often leads to increased wastewater treatment and maintenance costs, and requires additional cleaning of the garments to remove grit and broken stones from pockets (*2*). Cellulases have enabled the industry to produce a variety of fashion effects on the garments, while reducing the required treatment time (*7*). The indigo dye used in denim fabric dyeing does not penetrate deeply into the cotton fibers, therefore, cellulases can partially remove the dye from certain areas of the fabric by simply cleaving the loosely bound cellulose fibers from the fabric surface and the dye associated with them. Garment producers can generate a variety of abrasion effects by simply altering the process parameters, such as the ratio of water-to-stones, stones-to-fabric or fabric-to-water; the size or shape of the stones; enzyme treatment time, temperature and pH and enzyme loading, and by using various chemical additives (*2*).

In an enzymatic bio-stoning process, care must be taken to limit backstaining (redeposition of indigo molecules onto the fabric), and to ensure complete removal of cellulase residues from the fabric to prevent strength loss. Higher cellulase loadings in the bath can reduce backstaining by 25-40% (*6,8*). Commercial enzyme assay kits are available to monitor the presence of residual enzyme in garments and fabrics (Megazyme).

In addition to cellulases, a group of copper containing oxidative enzymes, known as laccases, have also been used in removing the indigo dye from denim fabrics (*9*). Campos *et al.* (*9*) demonstrated that purified laccases from *Trametes hirsuta* and *Sclerotium rolfsii* fungi were capable of degrading indigo dye to the intermediate compound known as isatin (indole-2,3-dione), and finally to anthranilic acid (2-aminobenzoic acid). By degrading the indigo dye molecules, laccases eliminate the problem of backstaining, and lower the concentration of dye molecules in the plant effluent. The other advantage of laccases is their lack of reactivity toward cellulose in cotton fibers.

Cellulases are also used to remove pills and fuzzy protrusion from the surface of cotton fabrics. Enzymatic depilling (biopolishing), can create smoother and softer fibers, improve pilling propensity, and enhance the drape and the durability of the fabrics by eliminating the use of strong chemicals (*10*). Cellulases, as well as xylanases and pectinases, have also been used to bio-polish non-cotton fibers, such as jute, Lyocell, rayon or wool fabrics (*10-12*). Cellulases are also added to household fabric softeners and detergents to enhance garment luster and appearance by removing the surface pills (*11*).

Unlike bio-stoning enzyme products, in which mono-component cellulases (endoglucanases) are used to limit the extent of cellulose degradation and subsequent fiber damage, the biopolishing enzyme products are multi-component cellulase preparations that contain endoglucanases, cellobiohydrolases, and β-glucosidases, that act synergistically to effectively hydrolyze the short protruding fibers known as pills (*12*).

Pectinases and Pectate Lyases in Bioscouring

To ensure proper dyeing and finishing, cotton fabrics are normally scoured using concentrated sodium hydroxide at high temperature and pH (caustic scouring) to remove hydrophobic components, such as fats, waxes, pectins and proteins (*13-14*). This process cannot only damage the fiber but is also highly energy and water intensive (*13,15-17*). Enzyme aided scouring (bioscouring), allows textile mills to operate their scouring units at lower temperatures and pH's, thereby reducing the post-scour rinse steps (*18-21*). Previous studies have shown that a hybrid scouring process (chemical-enzymatic) can produce white, highly absorbent cotton fabrics with improved dyeability (*17,22*).

Pectin comprises approximately 1% by weight of the whole cotton fiber, and is primarily composed of 85% methylated polygalacturonic acid backbone with arabinose and galactose side chains (*13,23*). Pectic substances are the major component of the middle lamella extracellular layer found between adjacent plant cells, which have been described as a "powerful biological glue" (*24*). Enzymes that degrade pectic substances are broadly named pectinases, and include polygalacturonases, pectin esterases, pectin lyases and pectate lyases (*13,25*). Pectin esterases, also known as pectin methylesterases, hydrolyze the functional ester group from the methylated acid groups. While pectinases cleave the unmethylated polygalacturonic backbone into galacturonic monomers, pectin lyases require methylated polygalacturonic monomers to perform the same reaction (*24,26*).

Not all bioscouring involves the use of pectin degrading enzymes. Csiszár *et al.* (*27*) observed that when desized fabric was treated with cellulase enzymes and then traditionally alkaline scoured, the processed fabric was significantly whiter and contained less seed-coat fragments as compared to controls (*27*). "Pretreatment" of the fabric with cellulase also allowed for lower hydrogen-peroxide consumption (*27*).

Other Applications of Enzymes in the Textile Industry

A variety of enzymes have been used to treat non-cotton fibers, such as wool, and jute and fiber blends such as jute-cotton and poly-cotton. Jute fibers are primarily comprised of hemicellulose, pectin and lignin (*28*). Cellulases can modify and partially hydrolyze jute fibers that protrude from the fabric surface, while pectinases and xylanases can degrade the pectin and xylan components, thereby softening the fabric (*10*).

Cellulases are also used to defibrillate Lyocell, a fiber manufactured and spun from wood pulp. The mechanical stress during the manufacturing process fibrillates the Lyocell fibers forming pill on the fabric surface. The traditional depilling process requires exposure of the fabric to soda ash at high temperatures, whereas biopolishing using cellulases can be accomplished under milder conditions and at the same time as desizing (*29*).

Proteases and Lipases for Wool Processing

Wool fibers are primarily composed of the fibrous protein, α-Keratin (97%) and lipids (1%), both of which can be partially hydrolyzed by proteases, lipases, and esterases. Current wool fiber modification technologies use strong halogenated chemicals, such as gaseous chlorine or chlorine salts, that lead to the generation of adsorbable organic halogens (AOX) in the effluents of textile wet processing plants. Enzymatic treatment using proteases is an effective and environmentally benign alternative that is currently being used in wool finishing operations to improve fiber dyeability, comfort (*e.g.*, softness and handle) and surface properties (*e.g.*, luster and drape). Protease treatments can also impart dimensional stability to wool fabrics and garments (shrink proofing). The mechanism by which proteases reduce the surface pill or prickles of the wool fiber is similar to the biopolishing of cotton by cellulases.

Felting or fiber shrinkage occurs when wool fibers are randomly mixed in water with or without mechanical agitation. The interlocking of the cuticle 'scales' on adjacent wool fibers causes these scales to move and contract toward the root end of the fiber (*30*). The outer surface of the cuticle, the epicuticle, contains both disulphide bonds and covalently bound fatty acids (25% by weight) that make this scaly formation highly hydrophobic (*31*). Work from the beginning of the last century showed that proteolytic enzymes can remove these scales and provide desirable anti-felting properties (*32*). However, the enzymatic finish has to be preceded by some type of chemical pretreatment (*e.g.*, pre-chlorination) to create some initial disruption in the cuticle surface, and increase the accessibility of the lipids and proteins to enzymes (*33*).

Catalases and Laccases for Mill Effluent Treatment

Textile wet processing effluents contain residual peroxide from bleaching operations, dyes, and other processing chemicals, and require extensive processing (*e.g.*, coagulation, precipitation, adsorption on activated carbon, ozone treatment, etc.) before the effluent can be either recycled or sewered. Catalase and laccases can be used to remove hydrogen peroxide and AZO-bonded dyes, respectively, from mill effluents and lower the cost and energy of the effluent treatment operations by up to 50% (*14,32,34-35*).

Catalases decompose the hydrogen peroxide to water and molecular oxygen at an extremely fast rate (500 x 10^6 mol H_2O_2/mol catalase/min) thereby reducing residence time in the wastewater treatment facility (*36*). To withstand the relatively intense conditions of the bleaching operation (60 °C and pH>9), catalases need alkaline and thermal tolerance properties. Paar *et al.* (*34*) identified several catalase and catalase-peroxidase enzymes from a novel strain of *Bacillus* isolated from textile bleaching effluents (*34*). The peroxide degrading ability of these enzymes under actual application conditions, *i.e.*, moderately high temperature and high pH, and in the presence of stabilizers and surfactants, has been extensively studied (*13,34,36-38*). From an application standpoint, the presence of stabilizers and surfactants has been shown to have a negative impact on both free and immobilized enzymes, reducing the enzyme activity by more than 90% (*e.g.*, Lutensit A-LB 50) (*34,36*). Immobilization of *Bacillus SF* catalase on alumina pellets improves the alkaline and thermal tolerance of the enzyme to a limited extent. Using a large horizontal packed-bed reactor containing immobilized catalase-peroxidase enzymes from *Bacillus SF*, Fruhwirth *et al.* (*37*) were able to reduce the water use in a dyeing operation by 45%.

Azo dyes used extensively by the textile industry have toxic, mutagenic, and carcinogenic properties and need to be thoroughly removed from mill effluents. Under anaerobic conditions, some bacteria can lower the concentration of Azo dyes by either metabolizing them or through biosorption (adsorption of dye onto cell biomass) (*35*). Laccases have been shown to remove chlorolignins, polycyclic aromatic hydrocarbons, and phenolic compounds from pulp mill and textile mill effluents (*39*). Abadulla *et al.* (*39*) found that enzyme immobilization is necessary to prevent the interference of enzymes with the dyeing process (*39*). Such interference may be caused by binding of the enzymes to the dye molecules, or aggregation of the dye molecules by the degradation products. Fungal laccases have been immobilized on a variety of carriers, such as alumina, activated carbon, agarose, Sepharose and porosity glass. The success rate of the immobilization technique for these enzymes is reported to vary between 70 to 98% with an enzyme activity recovery rate of 67 to 96%. Enzyme immobilization is known to improve enzyme stability and sensitivity to inhibitors.

Experimental

Enzyme Discovery

The source of pectinolytic genes for this study was DNA isolated directly from environmental soil samples; this DNA was used to make highly complex expression libraries (DiversaLibraries™). The methodology was developed to circumvent the inefficiency encountered in trying to grow all of the organisms present in environmental samples.

The model pectin substrate azo-rhamnogalacturonan was used to screen these environmental microbial genomic DNA libraries (*40*) to find novel pectinolytic enzymes with the potential to replace the chemicals currently used in the caustic scouring process. Newly-discovered pectate lyases were characterized and tested in a bioscouring application for optimal enzymatic properties, and compared to the conventional chemical scouring procedure. The best performing wild type enzyme was then evolved to deliver superior performance at higher temperatures using directed evolution approaches (*41-44*).

Optimization of the best wild type enzyme was undertaken by creating libraries of sequence variants, screening for improved enzyme characteristics, and then combining the beneficial sequence variations to create further improved variants. Initially, all possible single site mutants of the wild-type enzyme were created using GSSM™ technology (*43*). Following the identification of multiple beneficial single amino acid substitutions, a library representing all possible combinations of the mutations was created using Gene Reassembly™ technology, and screened to identify the optimal enzyme for this application. This enzyme was given the designation Cottonase™.

Scouring of Cotton Swatches at the Laboratory Scale

The 100% cotton fabric swatches (460R, Testfabrics, USA) were scoured in stainless steel beakers of a Mathis Labomat (Mathis, Concord, NC). Fabric was cut into 12 cm x 12 cm swatches (~2 g). Each swatch was weighed and, if necessary, trimmed until 2 g ± 0.02 swatches were obtained. Two swatches were used per beaker.

The required amount of buffer (50 mM carbonate/bicarbonate, pH 9) was placed in each beaker using a 10:1 liquor ratio based on the dry weight of two swatches of cotton fabric. Cottonase™ enzyme (for bioscouring) or sodium hydroxide (for caustic scouring), surfactant (Kieralon Jet B [BASF, USA] 0.75% on weight of fabric [owf]), chelant (Barapon C-108 [Dexter Chemicals, USA] at 0.5% owf), and control solutions (buffer) were added to the swatch-buffer mixture by placing pipettes partly into liquor solution, and dispensing and aspirating the pipette 3 times to ensure complete transfer of chemicals. Two fabric swatches were then loaded into each beaker, and the beaker was shaken to ensure proper wetting of the fabric. Caustic scoured swatches were used as the control; the caustic scouring solution contained 3% (w/w) NaOH and standard amounts of surfactant and chelants as described above.

The beakers were loaded into the Labomat and the temperature raised from 25 °C to 90 °C (4 °C/min) and held for 10 min with rotation set to 40 rpm (30 sec left, 30 sec right). The beakers were then removed, uncapped and the liquid was decanted. The swatches were rinsed with DI water by overflowing the beaker.

The swatches were then returned to their respective beakers and cold DI water was added (10:1 liquor ratio). Beakers were loaded into the Labomat and the heat treatment described above was repeated followed by overflow rinsing. Each swatch was marked using a Sanford Laundry Marking pen (Fisher Scientific) and neutralized by soaking briefly in DI water (pH adjusted to 7). One swatch from each beaker was used (wet) for Ruthenium Red staining, and the other replicate was dried and used for measuring the water absorbency by the Drop Test. Drying was undertaken in a laboratory oven at 70 °C for 3 hours, followed by conditioning the dry swatches at 20 °C and 65% relative humidity overnight.

Scouring and Dyeing of Cotton Fabric at the Pilot Scale

Pilot scale scouring was undertaken at North Carolina State University Textile Pilot Plant in a Werner Mathis JFO jet unit (NC, USA) using 1 kg of 100% Cotton fabric (Testfabrics, USA) and an 8:1 liquor ratio. A loop was formed by sewing the two ends of the fabric, and the loop was loaded onto the reel by passing the fabric through the venturi nozzle. Scouring chemicals (enzyme, surfactant, chelant, etc.) were mixed in the unit's dosing system and fed into the bath using pressurized hot water. The dosing containers were rinsed several times to ensure complete transfer of the scouring chemicals. The bath temperature was raised to 90 °C and held for 10 min. The bath was then emptied of liquor and the fabric was rinsed twice with fresh water.

The bioscoured fabrics were dyed with Remazol Turquoise G-A (Dystar, USA) at a 10:1 liquor ratio. The dye bath contained an antifoaming agent (Ciba Burst 2000 at 1.0 g/L), Remazol Turquoise G-A (2.0% on weight of fabric [owf]), sodium sulfate (50.0 g/L) and soda ash (8.0 g/L). The fabric was first loaded onto the jet reel and chemicals were added at 10 min intervals while circulating the bath. The bath temperature was raised to 80 °C (1 °C/min) and maintained at that temperature for 30 min with continuous circulation. The bath was then drained and fabric was rinsed by filling and draining the jet twice. To neutralize the fabric, acetic acid (1 g/L), fabric softener (Kieralon NF at 1 g/L) and water were added to the machine and heated to 70 °C circulating continuously for 10 min. Fabric was then rinsed twice with water, unloaded and dried. Uniformity of dyeing was examined by visual inspection and by measuring the color intensity spectrophotometrically (results not shown).

Assessing Scouring Efficiency

The efficiency of enzymatic bioscouring was measured using Ruthenium Red Stain which generates a reddish color by binding to the residual pectin left on the fabric. The fabric water absorbency, as an indication of its dyeability, was assessed using the standard Drop Test. Both methods are described below.

Residual Pectin Analysis by Ruthenium Red Staining

To prepare one liter of 0.05% (w/v) Ruthenium Red staining solution, 0.5 g of Ruthenium Red dye was added to 1 L of 50 mM sodium phosphate buffer (pH 6.0). The solution was then mixed for 15 min and filtered using a 0.2 μm filter.

To stain the scoured fabrics, single swatches were placed into Labomat beakers and 100 mL of 0.05% ruthenium red staining solution added to each beaker. Beakers were loaded into Labomat and temperature was raised to 50°C (6 °C/min) and held for 30 min. Beakers were removed and decanted, and each swatch was rinsed (overflow) with DI water and squeezed gently by hand. Swatches were dried at 50 °C for 3 hours. The luminosity of each swatch was determined using a ColorEye spectrophotometer (GretagMacBeth, New York, USA).

Water Absorbency Measurement (Drop Test)

The Drop Test was carried out according to AATCC Method 79 with minor modifications. Fabric swatches were tensioned horizontally on an embroidery ring and water drops of approximately 0.05 mL were dropped from a 40 mm distance. The time between the landing of a drop on the fabric and the disappearance of shine from the surface of the drop was measured. Five measurements were made per fabric swatch and the average absorbance time was determined.

Results and Discussion

The residual pectic polysaccharides content of treated fabrics was determined by staining the swatches with Ruthenium Red dye and measuring the luminosity of the stained fabric (Figure 1). Scouring with Cottonase™ (0.2 and 1% owf) effectively reduced the pectic polysaccharide (PPS) content of the fabric as demonstrated by the higher luminosity of the Cottonase™ treated swatches. The equivalent or greater luminosity of the Cottonase™ treated fabrics (0.2% and 1%) when compared with the caustic scoured swatch indicated that the enzymatic treatment was as effective as the industrially practiced caustic treatment, and produced fabrics suitable for dyeing. Cottonase at 0.05% was not as effective in removing PPS perhaps due to the limited amount of mechanical agitation at small scale. It is known that mechanical agitation is essential in order for chemicals or enzymes to be fully effective during various wet treatment operations (*45*). Some pectin was removed from swatches treated with only auxiliary chemicals. This was not surprising as the presence of surfactants at high temperature can dislodge hydrophobic surface material (*e.g.*, waxes) from the fabric, even in the absence of a pectin degrading agent, namely, enzyme or caustic solution.

In the textile dyeing and finishing industry, the rate at which a piece of fabric absorbs water droplets placed on its surface is used as a predictor for how well that fabric can be dyed. An unscoured piece of fabric usually has very limited absorbency for water, because the fiber surface of untreated cotton is covered with a matrix of pectic polysaccharides and hydrophobic substances that limit access to the more hydrophilic cellulose core of the fiber (*1, 13*). Treatment of fabric with pectin degrading enzymes such as Cottonase™, loosens this protective matrix and facilitates the removal of hydrophobic substances under milder conditions (*e.g.*, lower pH and short treatment time) without the need for harsh chemicals. It is, therefore, essential to demonstrate that the fabric has gained sufficient water absorbency upon completion of the scouring operations.

The hydrophilicity of the treated swatches was examined by the AATCC water drop test (Table I). The Cottonase™ bioscouring procedure was effective in enhancing the water absorbency of the fabric to ~3 sec. An absorbency time of 1-5 sec is considered acceptable by industry.

One interesting observation was the rapid water absorption (short absorbency time) for the auxiliary treated swatches where pectin was not effectively removed (Table I). This 'false' water absorbency could have been caused by the binding of highly hydrophilic surfactant molecules, used in the scouring bath, onto the fabric surface. A non-scoured fabric, with apparently

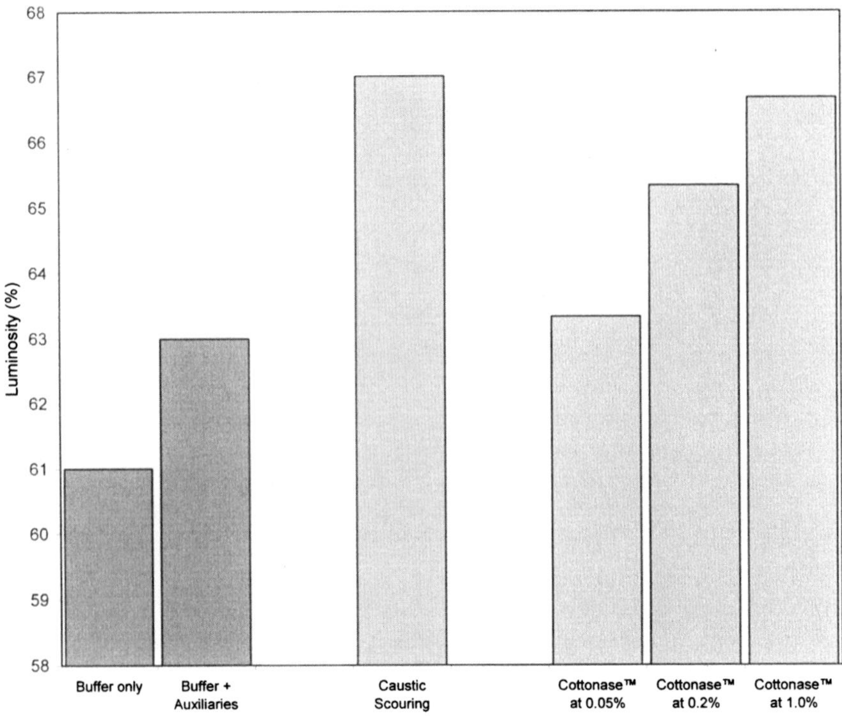

Figure 1. Luminosity of treated and stained (Ruthenium Red) Interlock cotton swatches. Higher luminosity indicates lower pectic polysaccharide content

Table I. Absorbency time of Interlock cotton swatches treated at lab scale

Treatment	*Absorbency time (sec)*
Bioscouring using Cottonase™ at:	
0.05% (owf)	4
0.2% (owf)	4
1.0% (owf)	5
Caustic scouring	1
Buffer + Auxiliaries	1
Buffer only	>>60

good absorbency, will not be a good candidate for dyeing, as the fibers still contain the waxy material that would limit interactions between cellulose and the dye molecules.

To verify the efficiency of Cottonase™ enzyme under simulated industrial conditions, kilogram quantities of fabric were scoured in a pilot scale jet type processing unit. Scouring was undertaken with both sodium hydroxide (caustic treatment) and the Cottonase™ enzyme product. By comparing the luminosity of fabrics scoured by enzyme or caustic, and stained with Ruthenium Red dye it was clear that Cottonase™ was able to effectively remove PPS from the fabric (Figure 2). This indicated that Cottonase™ at 0.1% and 0.2% (owf) was able to remove more PPS than that achieved by caustic treatment. No significant loss of fabric mass was observed after enzymatic treatment (data not shown).

Fabrics bioscoured at the pilot scale also had acceptable absorbency times (2-3 sec) indicating good dyeing properties (Table II). This was confirmed by dyeing the fabric with a turquoise dye. The uniformity and brightness of the dyed fabric passed industry accepted tests for quality (data not shown).

Conclusions

Cottonase™ bioscouring of 100% knit cotton fabric resulted in a highly absorbent fabric, and the Ruthenium Red staining of the scoured swatches

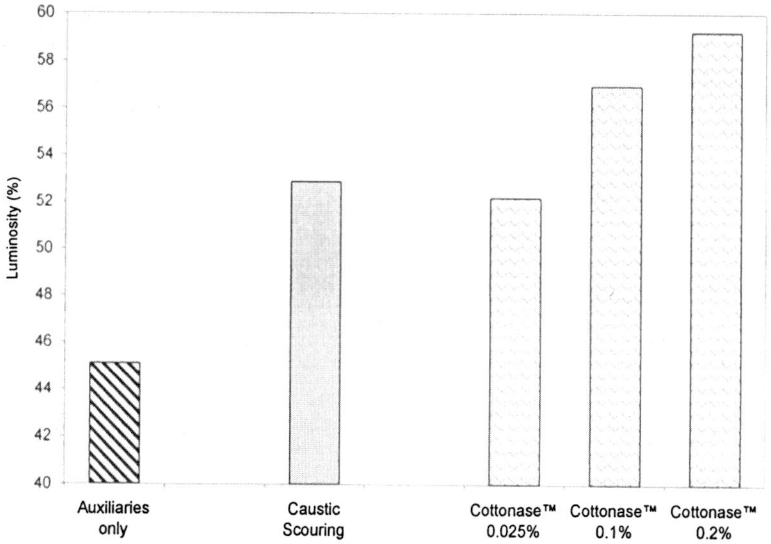

Figure 2. Luminosity of treated and stained (Ruthenium Red) Interlock cotton swatches. Higher luminosity indicates lower pectic polysaccharide content.

Table II. Absorbency time of Interlock cotton fabric treated at pilot scale

Treatment	*Absorbency time (sec)*
Bioscouring using Cottonase™ at:	
0.025%	2
0.1%	2
0.2%	3
Caustic scouring	6
Treatment with water + auxiliaries	>60

indicated that enzymatic treatment was as effective as caustic scouring in removing the pectic polysaccharides. The same observations were made when kilogram quantities of fabric were treated at pilot scale in a jet type processing unit. Fabrics bioscoured at pilot scale were successfully dyed, and the final material met the commercial standards for quality and dyeing uniformity. In summary, this study confirmed that enzymatic scouring with Cottonase™ is an effective alternative to conventional caustic scouring of cotton fabric.

References

1. Cavaco-Paulo, A.; Guebitz, G. M. In *Textile Processing with Enzymes*, Cavaco-Paulo, A.; Guebitz, G. M.; Eds.; **2003**, CRC Press: Boca Raton, p 86-119.
2. *AATCC Garment Wet Processing Technical Manual;* American Association of Textile Chemists and Colorists: Research Triangle park, NC, **1994**.
3. Gupta, R.; Paresh, G.; Mohapatra, H.; Goswami, V. K.; Chauhan, B. *Process Biochemistry* **2003**, *38*, 1599-1616.
4. Godfrey, T. In *Industrial Enzymology*, Godfrey, T.; West, S.; Eds.; Stockton Press: New York, Chapter 2.21, **1996**, p 359-371.
5. Deschler, O.; Schmidt, G. *Textile Praxis International,* **1981**, 9-11.
6. Mamma, D.; Kalantzi, S. A.; Christakopoulos, P. *Textile Chemist and Colorist and American Dyestuff Reporter* **2004**, *79*, 639-644.
7. Sariisik, M. *AATCC Review 4*, **2004**, 24-29.
8. Klahorst, S.; Kumar, A.; Mullins, M. M. *Textile Chemist and Colorist,* **1994**, *26*, 13-18.
9. Campos, R.; Kandelbauer, A.; Robra, K. H.; Cavaco-Paulo, A.; Guebitz, G. M. *J.Biotechnol.* **2001**, *89*, 131-139.
10. Sreenath, H. K.; Shah, A. B.; Yang, V. W.; Gharia, M. M.; Jeffries, T. W. *J. Ferment. Bioengineer.* **1996**, *81*, 18-20.
11. Chiweshe, A.; Cox-Crews, P. *Textile Chemist and Colorist and American Dyestuff Reporter* **2000**, *32*, 41-47.

12. Tyndall, M. I. *AATCC Review* **1992**, *24*, 23-26.
13. Hartzell, M. M.; Hsieh, Y. In *Enzyme Applications in Fiber Processing*, Eriksson K.L; Cavaco-Paulo, A.; Eds.; American Chemical Society: Washington, DC, Chapter 18, **1998**, p 212-227.
14. Heine E.; Hocker H. *Rev. Prog. Coloration* **1995**, *25*, 57-63.
15. Kirk, O.; Borchert, T. V.; Fuglsang, C. C. *Curr. Opin. Biotechnol.* **2002**, *13*, 345-351.
16. Sangwatanararoj, U.; Choonukulong, K.; Ueda, M. *AATCC Review* **2003**, *3*, 17-20.
17. Tzanov, T.; Silva, C. J.; Zille, A.; Oliveira, J.; Cavaco-Paulo, A. *Enzyme Microb.Technol.* **2001**, *29*, 357-362.
18. Choe, E. K.; Nam, C. W.; Kook, S. R.; Chung, C.; Cavaco-Paulo, A. *Biocatalysis and Biotransformation* **2004**, *22*, 375-382.
19. Liu, J.; Condon, B.; Showmaker, H. L. I. US B1 6544297, May 24, **2000**, pp 16-20030408.
20. Liu, J.; Salmon, S.; Kuildred, H. A. WO A1 2006002034, June 14, **2005**, pp 50-20060105.
21. Xu, H.; Liu, J.; Otto, E.; Condon, B. WO A1 2003002810, July 1, **2002**, pp 28-20030109.
22. Waddell, R. B. *AATCC Review* **2002**, *2*, 28-30.
23. Gamble G.R. *Textile Research Journal* **2003**, *73* (2), 157-160.
24. Hoondal, G. S.; Tiwari, R. P.; Tewari, R.; Dahiya, N.; Beg, Q. K. *Appl. Microbiol. Biotechnol.* **2002**, *59*, 409-418.
25. Alkorta, I. C. G. M. J. L. *Process Biochem.* **1998**, *33*, 21-28.
26. Solbak, A. I.; Richardson, T. H.; McCann, R. T.; Kline, K. A.; Bartnek, F.; Tomlinson, G.; Tan, X.; Parra-Gessert, L.; Frey, G. J.; Podar, M.; Luginbuhl, P.; Gray, K. A.; Mathur, E. J.; Robertson, D. E.; Burk, M. J.; Hazlewood, G. P.; Short, J. M.; Kerovuo, J. *J. Biol. Chem.* **2005**, *280*, 9431-9438.
27. Csiszár, E.; Szakacs, G.; Rusznák. In *Enzyme Applications in Fiber Processing*, Eriksson K.L; Cavaco-Paulo, A.; Eds.; American Chemical Society: Washington, DC, **1998**, Chapter 17, 204-211.
28. Macmillan, W. G.; Sengupta, A.; Roy, A. *J. Ind. Chem. Soc.* **1955**, *32*, *80-83*.
29. *Lyocell just needs salts and enzymes.* http://www.novozymes.com/cgibin/bvisapi.dll/biotimes/one_article.jsp?id=11479&lang=en, **1999**.
30. Galante, Y. M.; Foglietti, D.; Tonin, C.; Innocenti, R.; Ferrero, F.; Monteverdi, R. In E*nzyme Applications in Fiber Processing*, Eriksson K.L; Cavaco-Paulo, A.; Eds.; American Chemical Society: Washington, DC, Chapter 24, **1998**, 294-305.
31. Kadolph, S. J.; Langford, A. L. In *Textiles*, Prentice Hall: Upper Saddle River, NJ. Chapter 5, **2002**.

32. Guebitz, G. M.; Cavaco-Paulo, A. *J. Biotechnol.* **2001**, *89*, 89-90.
33. Heine, E.; Hollfelder, B.; Lorenz, W.; Thomas, H.; Wortmann, G.; Hoecker, H. In *Enzyme Applications in Fiber Processing*, Eriksson K.L; Cavaco-Paulo, A.; Eds.; American Chemical Society: Washington, DC, **1998**, p 279-293.
34. Paar, A.; Costa, S.; Tzanov, T.; Gudelj, M.; Robra, K. H.; Cavaco-Paulo, A.; Guebitz, G. M. *J. Biotechnol.* **2001**, *89*, 147-153.
35. Chen, K. C.; Huang, W. T.; Wu, J. Y.; Houng, J. Y. *J. Ind. Microbiol. Biotechnol.* **1999**, *23*, 686-690.
36. Costa, S. A.; Tzanov, T.; Paar, A.; Gudelj, M.; Guebitz, G. M.; Cavaco-Paulo, A.. *Enzyme Microb.Technol.* **2001**, *28*, 815-819.
37. Fruhwirth, G. O.; Paar, A.; Gudelj, M.; Cavaco-Paulo, A.; Robra, K. H.; Guebitz, G. M. *Appl. Microbiol. Biotechnol.* **2002**, *60*, 313-319.
38. Gudelj, M.; Fruhwirth, G. O.; Paar, A.; Lottspeich, F.; Robra, K. H.; Cavaco-Paulo, A.; Guebitz, G. M. *Extremophiles.* **2001**, *5*, 423-429.
39. Abadulla, E.; Tzanov, T.; Costa, S.; Robra, K. H.; Cavaco-Paulo, A.; Guebitz, G. M. *Appl. Environ. Microbiol.* **2001**, *66*, 3357-3362.
40. Short, J. M. *Nat. Biotechnol.* **1997**, *15*, 1322-1323.
41. Palackal, N.; Brennan, Y.; Callen, W. N.; Dupree, P.; Frey, G.; Goubet, F.; Hazlewood, G. P.; Healey, S.; Kang, Y. E.; Kretz, K. A.; Lee, E.; Tan, X.; Tomlinson, G. L.; Verruto, J.; Wong, V. W.; Mathur, E. J.; Short, J. M.; Robertson, D. E.; Steer, B. A. *Protein Sci.* **2004**, *13*, 494-503.
42. Gray, K. A.; Richardson, T. H.; Kretz, K. A.; Short, J. M.; Bartnek, F.; Knowles, R.; Kan, L.; Swanson, P.; Robertson, D. E. *Advanced Synthesis and Catalysis* **2001**, *343,* 607-617.
43. Kretz, K. A.; Gray, K. A.; Robertson, D. E.; Short, J. M. *Methods in Enzymology* **2004**, 388, 3-11.
44. Short, J. M. U.S. 6,537,776, **2003**.
45. Cavaco-Paulo, A. In *Enzyme Applications in Fiber Processing*, Eriksson, K. L.; Cavaco-Paulo, A.; Eds.; American Chemical Society: Washington, DC, **1998**, p 180-189.

Chapter 10

Intensification of Enzymatic Reactions in Heterogeneous Systems by Low Intensity, Uniform Sonication: New Road to "Green Chemistry"

Val G. Yachmenev, Brian D. Condon, and Allan H. Lambert

Cotton Chemistry and Utilization Research Unit, Southern Regional Research Center, Agricultural Research Service, U.S. Department of Agriculture, New Orleans, LA 70124

Enzymatic bio-processing of various substrates generates significantly less wastewater effluents that are readily biodegradable, and do not pose an environmental threat. However, two major shortcomings that impede acceptance by industry are expensive processing costs and slow reaction rates. Enzymatic bio-processing of cotton textiles with cellulase and pectinase enzyme processing solutions, under low energy, uniform sonication greatly improved the efficiency of the enzymes and significantly increased the overall reaction rates. Specific critical features of combined enzyme/ultrasound bio-processing are: a) cavitation effects that enhance the transport of enzyme macromolecules toward the substrate surface, b) the several hundred fold greater effect of cavitation in heterogeneous than homogeneous systems, and c) the maximum effects of cavitation at ~50 °C that are optimum temperature for most enzymatic reactions. Contrary to common belief, low intensity, uniform sonication does not damage or inactivate sensitive structures of enzyme protein macromolecules. Overall, under specific conditions, carefully controlled introduction of ultrasound energy during enzymatic bio-processing has an excellent potential for the intensification of a variety of technological processes that involve many types of industrial enzymes and substrates.

Introduction

Since the middle 1990s, the use of various enzymes in the textile industry has increased significantly, especially in the processing of natural fibers such as cotton. A major incentive for embracing enzymatic bio-processing was that the application of enzymes is much more environmentally benign, and the reactions catalyzed are very specific and ensure a more focused performance. In contrast, traditional chemical processing is much less specific and often results in undesirable side effects, *e.g.*, significant reduction in the degree of polymerization of cellulose. Enzymes used in cotton bio-processing act as catalysts to speed up complex chemical reactions such as the hydrolysis of cellulose by cellulases, pectins by pectinases, starches by amylases, and triglyceride-based compounds in fats and oils by lipases. As enzymes are catalysts, relatively small concentrations are required. If conditions are favorable to the specific enzyme, it will repeat the action (hydrolysis) many times during the process. Other potential benefits of enzymatic bio-processing include cost reduction through energy and water savings, and improved product quality. Even greater acceptance of enzymatic bio-processing by the textile industry in the near future will probably result from increasing legislative pressures by governments worldwide to sharply decrease the quantity and toxicity of textile wastewater effluents. The typical applications of enzymes for bio-processing of cotton are summarized in Table I (*1-4*).

Although there are numerous advantages of enzymatic bio-processing of cotton fibers there are several shortcomings such as added processing costs and relatively slow reaction rates. Enzymatic processing of cotton textiles, like any wet processing system, involves the transfer of mass (enzyme macromolecules) from the processing liquid medium (enzyme solution) across the surface of the textile substrate. The detailed mechanism of enzymatic reactions is quite complicated and is still being investigated. In general terms, the enzymatic reaction could be summarized according to the stages illustrated in Figure 1. At least two stages of enzymatic reaction (Figure 1) involve transport of the enzyme macromolecules and enzymatic reaction products to and from the fiber surface. Since both stages are controlled by diffusion, the overall reaction rate of enzymatic hydrolysis is governed by the diffusion rate of the enzyme macromolecules. In general, large three-dimensional enzyme macromolecules have low diffusion rates and tend to react with outlaying cellulose fibers in cotton yarn, which could result in excessive fiber damage.

It has previously been suggested by the authors (*5-6*) that sonication of the enzyme processing solution under certain specific conditions could provide a much more efficient transport mechanism for "bulky" enzyme macromolecules throughout the immediate border layer of liquid at the substrate surface.

Table I. Typical Examples of Enzymes Used in Cotton Textile Processing

Application	*Enzyme(s)*	*Benefit*
Desizing of cotton	Amylase	Removal of starch from the surface of fiber
Scouring of greige cotton	Pectinases, cellulases, lipases	Removal of waxes, proteins, pectins and natural fats from the surface of cotton fibers
Peroxide breakdown	Catalase	Effluent treatment to remove residual H_2O_2
Bio-finishing of cotton	Cellulases	Improvement of the appearance of cotton fabrics and garments by removal of fiber fuzz and pills from the substrate surface
Bio-stoning of denim	Cellulases	"Stone-washing" of denim fabrics to produce the fashionable aged appearance
Bio-bleaching of denim	Laccases	"Stone-washing" effects without loss of fabric strength
Laundry washing	Proprietary mixtures of enzymes	Removal of soils and stains

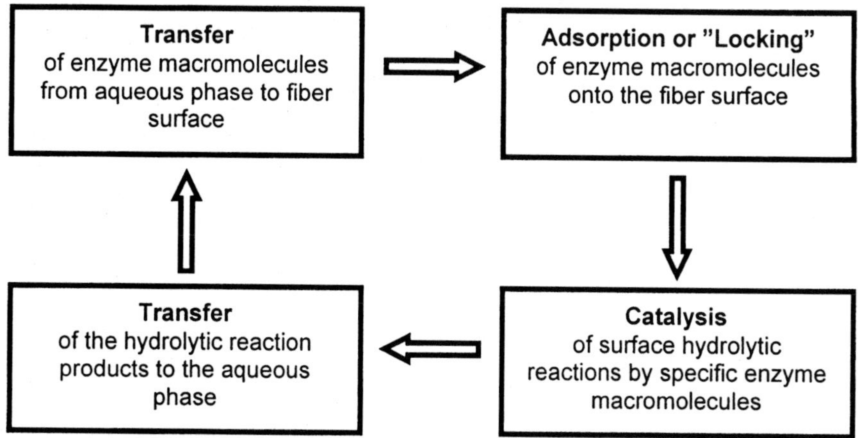

Figure 1. Schematic diagram of the general stages of an enzymatic reaction on a solid substrate.

Technical Aspects of Application of Ultrasound for Intensification of Enzymatic Bio-processing

Generally, introduction of ultrasound energy into the liquid medium has two primary effects: cavitation and heating (Figure 2). In case of enzymatic bio-processing, the more important of these two is cavitation - formation, growth, and implosive collapse of bubbles in a liquid. The dynamics of cavity growth and collapse are highly dependent on the type of liquid, presence of dissolved species and gases in liquid, and temperature of liquid. The imploding cavitation bubble (Figure 2; bottom; a high-speed flash photomicrograph [7]) causes the nearly adiabatic compression of the excess vapors inside of the cavity thus raising its pressures ~ 500 atm. and temperatures ~ 5,500 °C (plasma conditions). It is important, that the sonication of liquid by low frequencies dissipates most of ultrasound energy through cavitation phenomena, while sonication by high frequencies dissipates significant amount of energy through heating (at the expense of cavitational dissipation).

The powerful ultrasonic irradiation of liquids produces a plethora of high energy chemical reactions that have been studied for many years. As the excess vapors in the cavitational bubble are compressed by the collapse, and its contents reach several thousand Kelvin, the trapped vapors are largely dissociated. For water, collapse of cavitational bubble produces the high-energy intermediates such as **H•** (atomic hydrogen), **OH•** (hydroxyl), $\mathbf{e^{-}_{(aq)}}$ (solvated electrons), $\mathbf{H_2O_2}$ (hydrogen peroxide), $\mathbf{HO_2}$ (superoxide) and, $\mathbf{H_2}$ (molecular hydrogen). The distinctive brand of chemistry - sonochemistry specifically studies the reactive interactions of these high-energy intermediates with various dissolved species in liquid (*8*).

$$\begin{aligned}
&O_2 \rightarrow 2\,O\bullet \\
&H_2O \rightarrow H\bullet + OH\bullet \\
&OH\bullet + OH\bullet \rightarrow H_2O_2 \\
&2\,O\bullet + H\bullet \rightarrow HO_2 \\
&H\bullet + H\bullet \rightarrow H_2
\end{aligned} \tag{1}$$

The formation of such highly reactive intermediates by collapsing cavitational bubbles should significantly affect the long term catalytic stability and activity of dissolved enzyme macromolecules. The common perception was that these highly reactive intermediates, and powerful shock waves resulting from collapse of cavitation bubbles, could severely damage or at least inactivate the very sensitive and intricate structures of enzyme proteins. However, when sonication was specifically used to inactivate enzymes and terminate enzymatic activity its efficiency was low (*9*). For example, it was reported that combined effects of heat, ultrasonic waves, and pressure were used for inactivation of certain thermostable enzymes (*10*).

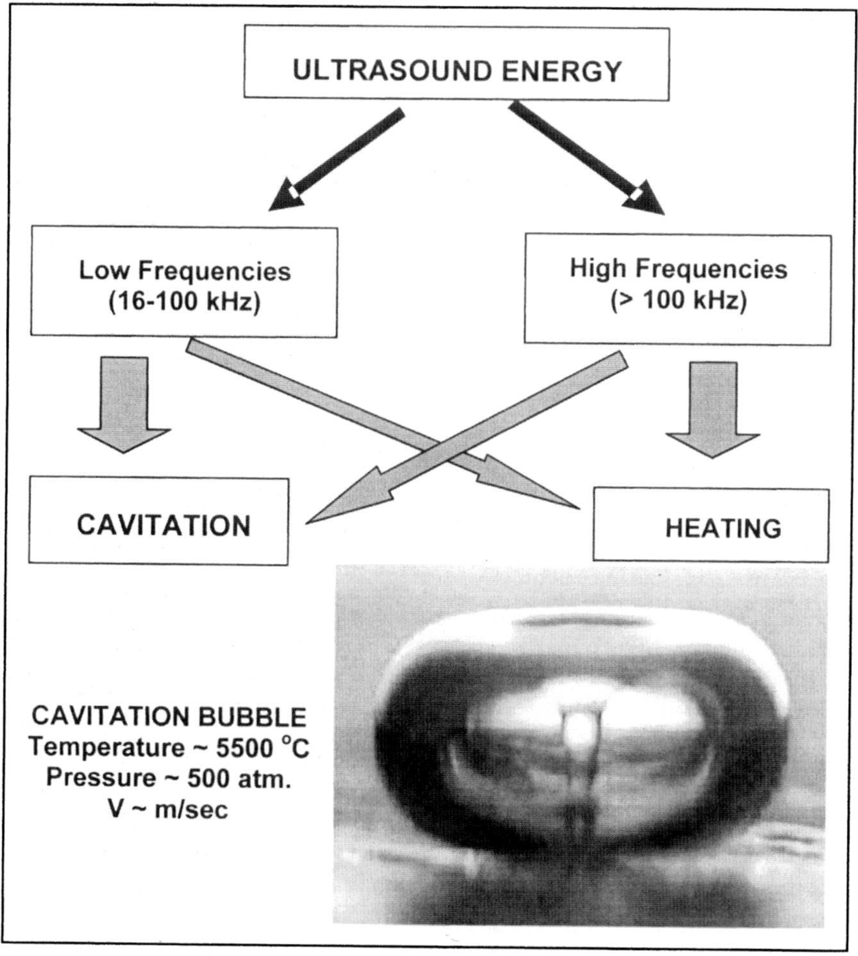

Figure 2. Schematic diagram of basic properties of ultrasound

The authors showed that synergistic effects of mano-thermosonication could, to a certain degree, reduce enzyme resistance to thermal inactivation. In another comprehensive overview of the combined effects of heat, pressure and ultrasound on microorganisms and enzymes (*11*), authors concluded that the resistance of most microorganisms and enzymes to ultrasound is so high that the required intensity of ultrasound treatment would be impractical. One possible explanation of the apparent inefficiency of ultrasound to inactivate enzyme macromolecules could be their extremely low ratio to the huge quantity of solvent molecules (*e.g.*, water) at the typically used enzyme processing concentrations of 4-5 g/liter. Therefore, the probability of enzyme macromolecules to be seized into a cavitation bubble, and to encounter the highly reactive intermediates created by collapsing bubbles should be very low.

If ultrasound, as it appears from the literature, does not affect the *specific activity* of industrial enzymes in any significant way, it could be used for intensification of enzymatic processing of cellulosic-based substrates by improving the *transport* of enzyme macromolecules towards the substrate's surface. Unlike the cavitation bubbles collapse in homogenous systems (liquid-liquid interface), collapse of a cavitation bubbles in heterogeneous systems (*e.g.*, enzyme solution - substrate) is different, because cavitation bubbles that collapse on or near to a surface are non-symmetrical as the surface provides resistance to liquid flow. The result is an in-rush of liquid predominantly from the opposite side of the bubble (remote from the substrate surface) resulting in a powerful liquid jet (roughly 450 m/sec) being formed and targeted at the surface. Because of the reduced liquid tensile strength at the liquid-solid interface, lower sonication intensities can be used in heterogeneous systems.

It is also important, that the rapid collapse of the cavitation bubbles generates significant shear forces in the bulk liquid immediately surrounding the bubble, to produce a strong stirring mechanical effect. This can significantly increase mass and heat transfer to the surface of the substrate by disrupting the interfacial boundary layers and also activating the catalytic performance of the enzyme macromolecules adsorbed onto the surface of substrate.

The diffusion transport of enzyme macromolecules toward the surface of a solid substrate could be also enhanced somewhat by simple mechanical agitation of processing solution. However, it is well known that mechanical agitation is not an effective stirring mechanism for the immediate border layer of liquid at a solid - liquid interface where the enzymatic reaction actually occurs (*12*).

Figure 3 presents the schematic distribution of the velocities in the layers of liquid concentrically surrounding the solid particle (substrate). The first, immediate layer of liquid at solid - liquid interface is motionless, and then velocities of following layers quickly increase to the maximum constant value defined by the power of agitation of the bulk solution. Since the immediate adjusted layer of liquid at a solid – liquid interface is practically immobile the only available transport mechanism for enzyme macromolecules to reach substrate surface is diffusion, which in case of such large protein macromolecules (50 000 - 250 000 Da) is not very efficient.

When microscopic cavitation bubbles collapse in the immediate vicinity of a substrate surface, they generate powerful shock waves that cause effective stirring/mixing of this adjusted layer of liquid. These shock waves generated by cavitation bubbles collapsing on and near the surface of substrate (*e.g.*, cellulosic fibers) are an ideal stirring mechanism for the immediate layer of liquid at the solid-liquid interface where enzyme reactions takes place (Figure 4).

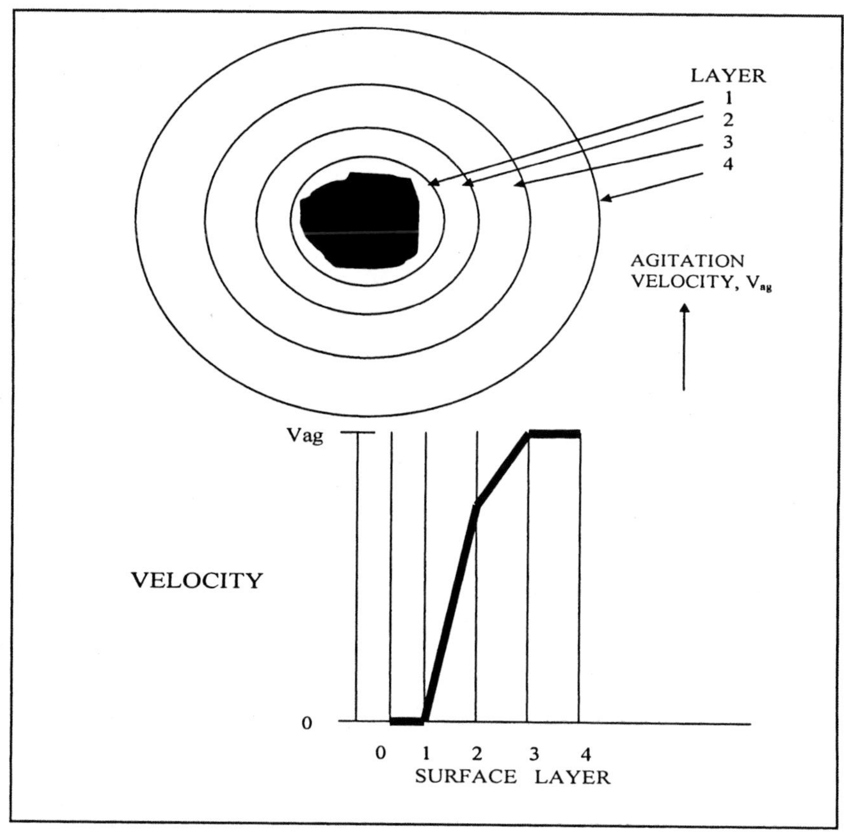

Figure 3. Schematic distribution of the velocities of the layers of liquid concentrically surrounding the solid particle (substrate).

The forceful stirring/mixing of this normally immobile layer of liquid greatly improves supply of enzyme macromolecules to the surface of a substrate. Following specific features of cavitation phenomena are very important for practical bio-processing application: a) effect of cavitation is several hundred times greater in heterogeneous (*e.g.*, all textile wet processes) than in homogeneous systems and b) in water, maximum effects of cavitation occur at ~50 °C, which is the nearly optimum temperature for most enzymatic bio-processing applications (*7*).

It is essential that the *uniform* introduction of ultrasound energy into heterogeneous systems will generate the majority of cavitation bubbles in the immediate vicinity of the solid-liquid interface because of asymmetry of surface tension, while in the case of homogeneous systems, cavitation bubbles are distributed evenly throughout the bulk of the processing solution.

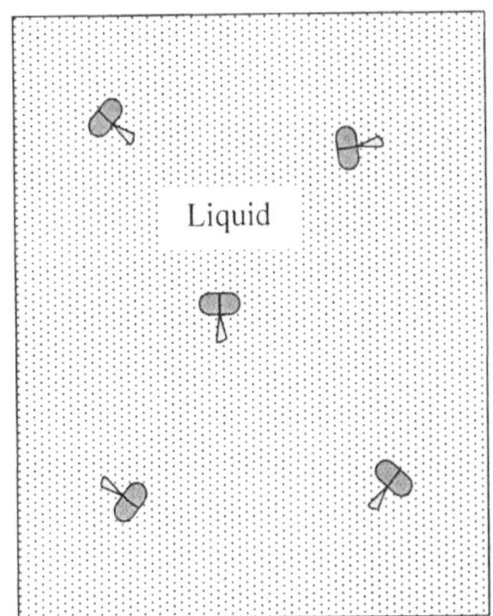

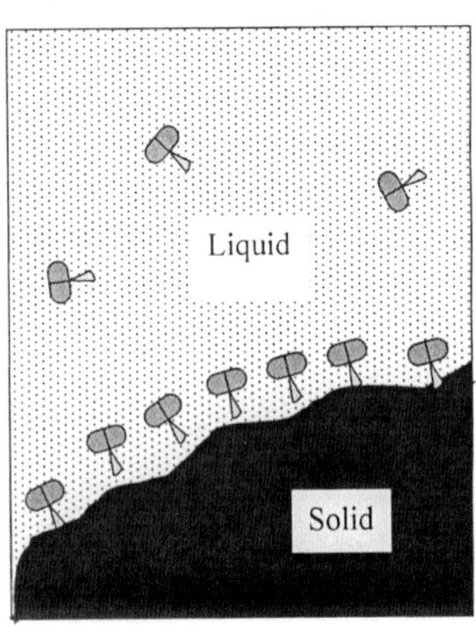

Figure 4. Distribution of collapsing cavitation bubbles in homogenous systems (liquid-liquid interface) and heterogeneous systems (solid-liquid interface).

It is also important that in case of heterogeneous systems most of cavitation bubbles are generated close to the surface of substrate thus providing an important additional benefit of the "*opening up*" of the surface of solid substrates as result of mechanical impacts produced by collapsing cavitational bubbles.

Another imperative consideration is that that despite their close-packed and generally well-ordered structures, enzyme macromolecules are usually not entirely rigid and have some conformational flexibility in solution that helps them to properly position the "active" domain relative to the substrate. Therefore, vigorous stirring/agitation of the normally immobile border layer of liquid at the liquid-solid interface caused by sonication should help enzyme macromolecules more easily position themselves "*fittingly*" onto the substrate (Figure 5).

And lastly, another valuable benefit of the intensive stirring/agitation of this border layer by collapsing cavitation bubbles is enhanced removal of the products of hydrolysis reaction from the reaction zone that should contribute to an overall increase in the reaction rate.

In summary, we can formulate the necessary requirements to maximize the benefits of ultrasound energy for enzymatic bio-processing as follows:

- **Ultrasound Frequency:** it appears that optimum sonication frequency should be in range of 20—100 kHz. These low sonication frequencies are more beneficial because:

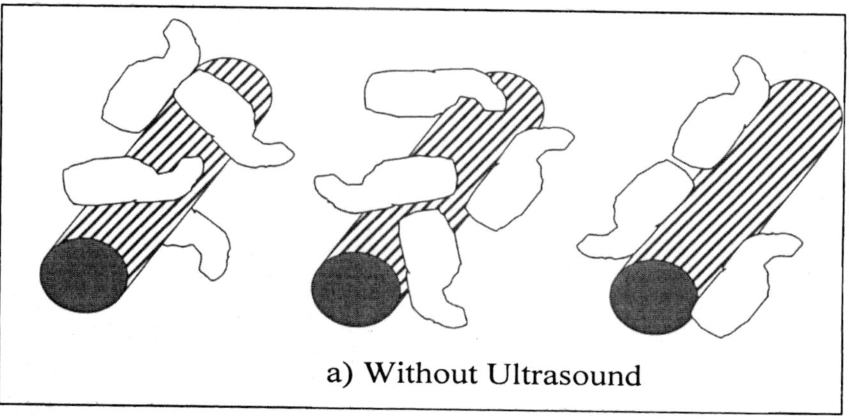

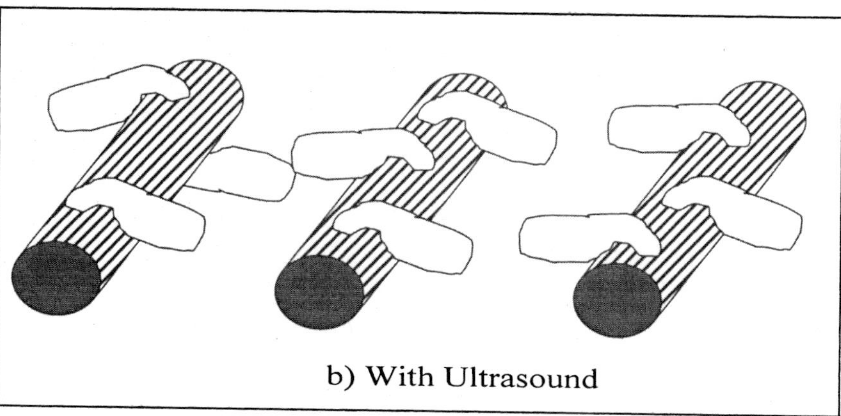

Figure 5. Scheme of the interaction of enzyme macromolecules with substrate with/without ultrasound.

a) Most of the introduced ultrasound energy is dissipated through the cavitation mechanism rather than through heating.
b) The lower sonication frequencies produce larger cavitation bubbles and therefore, more powerful "jets", thus providing more vigorous stirring/mixing of the border layer of liquid at solid - liquid interface.
c) Equipment for generation of low sonication frequencies is more cost efficient because of use of more durable, less expensive magnitostrictive transducers.

- **Ultrasound Energy:** optimum sonication power should be in range of 2-10 W/cm^3. The low energy sonication of enzyme processing solution improves the transport of enzyme macromolecules, but does not generate excessive amount of highly reactive intermediates.

- **Uniformity of Introduction of Ultrasound Energy:** it is critical that ultrasound energy is introduced into the processing bath in the most uniform way. This ensures the uniform generation of cavitation bubbles throughout the bath and uniformity of enhancement of transport of enzyme macromolecules toward substrate. Interestingly, the example of the probably most unsuitable sonication device to study the effects of ultrasound on enzymes performance is the typical unit shown at Figure 6 (left) and yet it is mostly used in laboratories. Figure 6 (right) shows that most of 1000 W of energy of this unit dissipates in ~1-3 cm^3 of solution (under the tip) and as result effectively obliterates enzyme macromolecules.

- **Application of Ultrasound in Heterogeneous vs. Homogeneous Systems:** since the effects of cavitation are several hundred times greater in heterogeneous than in homogenous systems, introduction of ultrasound could be economically justified only for solid-liquid systems. In homogenous systems, the much less expensive mechanical agitation will probably suffice.

There are a number of papers on the application of ultrasound energy mostly for the intensification of some conventional textile processing such as desizing, peroxide bleaching, and dyeing of natural fibers (*13-14*). Despite the apparent attractiveness of the introduction of ultrasound energy for intensification of enzymatic bio-processing of natural fibers, it was unclear how ultrasound affects complex structures of enzyme macromolecules, and how significant the benefits of introduction of ultrasound energy were.

The objectives of our experiments were to study the influence of low level, uniform sonication on the enzymatic bio-scouring of greige cotton with commercial BioPrep L^{TM} pectinase at a variety of concentrations/treatment times, and also the influence of ultrasound on the enzymatic bio-finishing of various types of cotton textiles with Cellusoft L^{TM} enzyme.

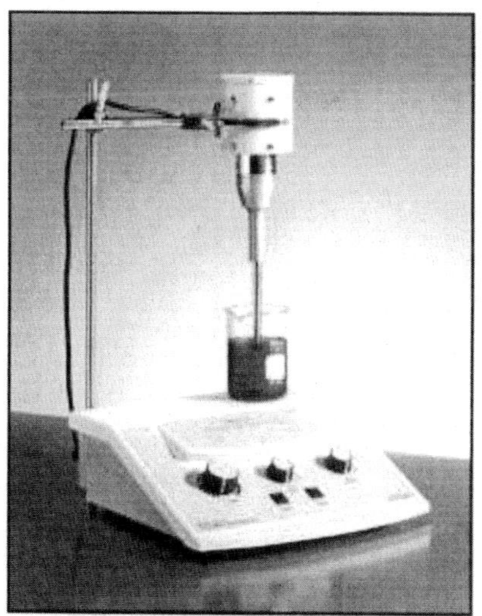

Figure 6. Typical laboratory ultrasound sonicator.

Experimental Ultrasonic Reactor

The enzymatic bio-processing of all textile samples was conducted in a parallel-plate Near-Field Acoustical Processor (NAP) (Figure 7). This dual frequency ultrasonic reactor with thermal control capabilities was specially designed and manufactured by Advanced Sonic Processing Systems Corporation for controlled sonication of textile samples (*15*). The ultrasonic reactor accommodates a single stainless steel frame with the sample precisely placed between two opposing diaphragm plates, which form two walls of the reaction chamber of the NAP. These diaphragms, driven by two sets of transducers (16 kHz and 20 kHz) produce a *beat frequency* inside the reaction chamber. This *beat frequency* develops a highly energized zone that continuously oscillates within the reaction chamber and insures consistent processing. Both ultrasound generators (1.1 kW) feeding the ultrasound transducers were equipped with variable power features allowing precise control of the intensity of ultrasonic energy inside the reaction chamber in the range of 0 – 3 W/cm^2.

The uniformity of sample's sonication in this unit was somewhat adequate but still far from ideal. An aluminum foil sample test indicated that only slightly

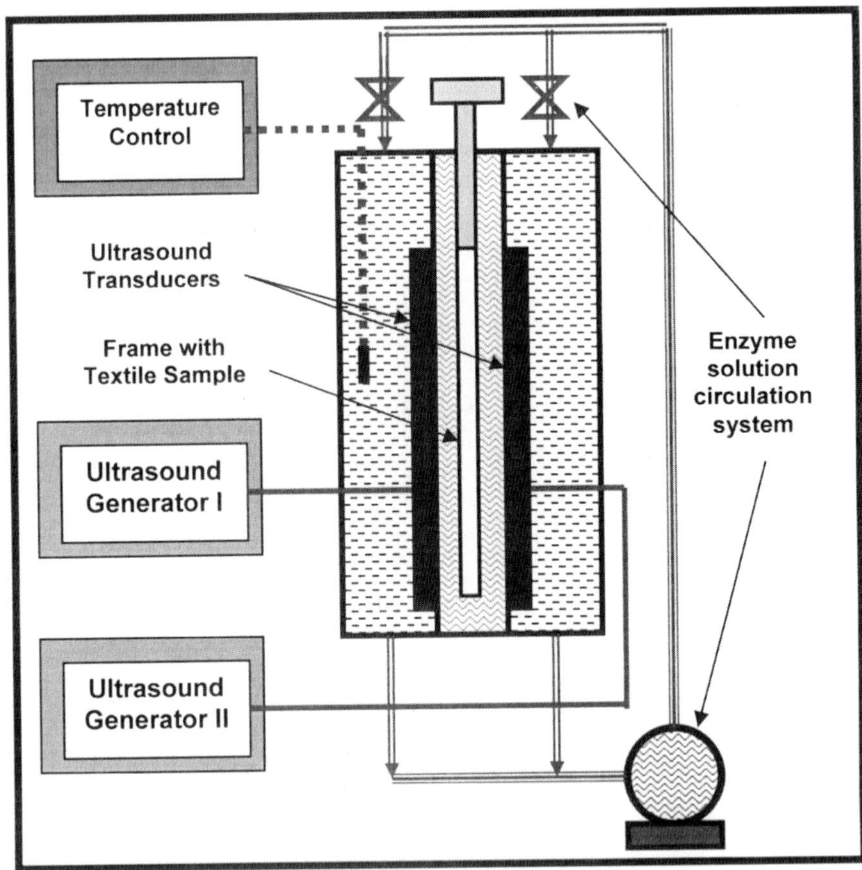

Figure 7. Schematic diagram of the experimental cell for combined enzyme/ultrasonic treatment of textile samples.

more than half of the total surface of the sample was covered with pinholes punched by collapsing cavitation bubbles. This was probably because the active surface of diaphragm plates with ultrasound transducers was only slightly more than 50% of the sample surface.

Examples of the Application of Sonication for the Intensification of Enzymatic Bio-processing of Cotton Textiles

Materials

Desized, scoured, bleached, and mercerized clean cotton printcloth, batiste, and broadcloth were used for all bio-finishing tests, and original greige cotton printcloth was used in all bio-scouring tests. Cotton fabrics were Testfabrics, Inc. Fabric samples (508 mm x 330 mm; ~25 g) sewn around the edges to prevent unraveling during processing. At least three samples of the same fabric were treated under identical conditions for each trial to ensure reproducibility. In bio-finishing tests, samples were dried at 110 °C for 6 h in weigh bottles to determine dry weight before and after enzyme/ultrasonic treatment. The weight loss, wettability, and breaking strength for all samples treated under various conditions with enzyme and/or ultrasound were determined and compared with samples of the original untreated cotton fabric. All fabric measurements were performed under constant conditions of 21 °C and 65% humidity. The wettability of treated samples was evaluated in accordance with a new AATCC Wicking Test (based on measurement of the time during which water is wicked up a distance of 3 cm on a strip of tested fabric in the warp and fill directions). Whole cellulase enzyme (Cellusoft L) ™ and alkaline pectinase (BioPrep L)™, from Novo Nordisk BioChem Inc., were used for the bio-finishing and bio-scouring tests. Enzyme assay and reaction conditions are listed in Table II.

Study Example I. Pectinase Bio-scouring Tests

Raw unscoured (greige) cotton contains ~90% cellulose and various non-cellulosics such as waxes, pectins, proteins, fats, and coloring matter. To remove these hydrophobic non-cellulosics and produce a highly absorbent fiber that can be dyed and finished uniformly, the greige cotton is traditionally processed with boiling sodium hydroxide solution in the presence of wetting and sequestering agents (*16*). This industrial process requires large quantities of water and energy, and generates a highly alkaline wastewater effluent. Alkaline pectinases might be a valuable alternative to harsh alkaline solutions in the preparation of cotton. At present, enzymatic bio-preparation of greige cotton represents a new approach and is mostly in the developmental stage (*17*).

Table II. Enzyme Assay and Reaction Conditions.

Enzyme	*Activity U/g*	*Buffer*	*pH*	*Temp. °C*
Cellusoft L	1500 ncu/g	Acetate, 0.01 M.	4.9	50
BioPrep L	3000 apsu/g	Phosphate, 0.01 M.	8.2	55

Two bio-preparation tests were undertaken with identical conditions of temperature, pH, sonication power, and circulation rate of enzyme processing solution through the reaction chamber of the acoustical processor, but with variable concentrations of BioPrep L™ alkaline pectinase (**test 1**) and treatment time (**test 2**). For each test, three samples of greige cotton printcloth were treated with BioPrep L solution with/without sonication (Liquor Ratio = 200:1). The primary objective of **test 1** was to determine how enzymatic bio-scouring of cotton printcloth with various concentrations of pectinase enzyme (0.1-1.0 g/L) would be affected by sonication. Treatment time of 30 min was selected primarily to emphasize the possible effect of sonication on the enzymatic bio-scouring rather than to achieve maximum absorbency of the sample. The average wettabilities of original samples and samples treated with various concentrations of BioPrep L pectinase with/without ultrasound are presented in Figure 8.

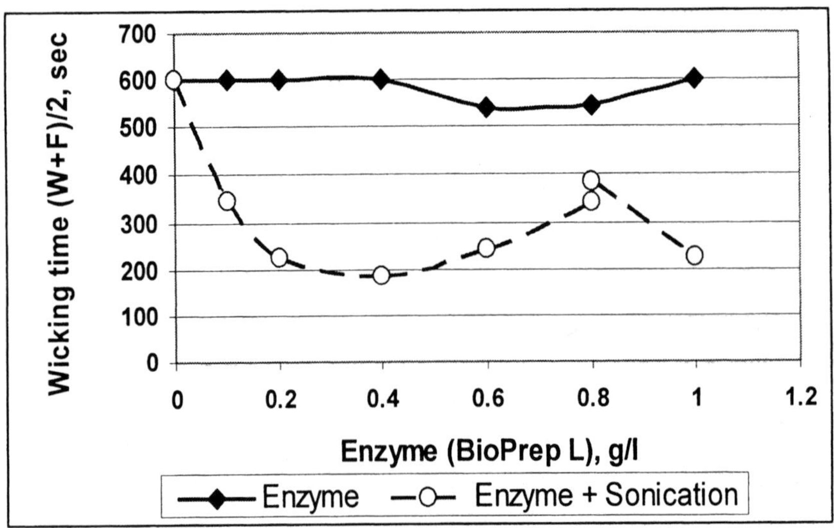

Figure 8. Evaluation of average wettabilities (Warp + Fill)/2 of cotton printcloth samples after alkaline pectinase bio-scouring under sonication conditions (test 1).

Experimental data on the wettability of all treated samples clearly showed that the introduction of ultrasound energy during pectinase bio-scouring greatly accelerated the process. The maximum effect provided by sonication occurs at a concentration of 0.4 g/L, which is lower than recommended by the manufacturer (0.5 g/L). The overall uniformity of bio-preparation was substantially better with combined pectinase/ultrasound treatment than with pectinase only treatment. Also, the benefit provided by sonication was less at the concentration of BioPrep L 0.8 g/L. To verify this atypical data, a test with that concentration was repeated. A possible explanation of the observed phenomena is that at such a relatively high concentration some enzyme macromolecules adsorb "non-fittingly" onto the surface of the fiber, "trapping" the waxes and pectins underneath and preventing their removal.

The objective of **test 2** was to investigate the influence of treatment time on the bio-scouring performance of BioPrep L™ alkaline pectinase under sonication conditions. For **test 2** the concentration of BioPrep L was also lowered to make the effect of sonication on enzymatic bio-scouring more pronounced. Average wettabilities of triplicate samples treated with BioPrep L pectinase with/without ultrasound are presented in Figure 9.

The introduction of ultrasound energy during pectinase bio-scouring again greatly accelerated the process. The maximum effect provided by sonication occurred at relatively short treatment times: 15, 30, 45, and 60 min. Trials with identical treatment time, but without sonication, showed poor overall absorbency; wicking times were more than 600 sec for 15, 30, and 45 min and 425 sec at the 60 min trial. At increased treatment times, *i.e.*, >90 min, sonication provided only modest improvement in pectinase performance.

Study Example II. Enzymatic Bio-finishing of Cotton Textiles

In bio-finishing applications, certain types of cellulase enzymes are used to impart an aged (denim) or renewed look on cotton fabrics. The process of renewal, known sometimes as bio-polishing or de-pilling, is based on the removal of fuzz and pills to give an improved surface appearance, which can be described as:

- Cleaner surface with a cooler feel
- Increased softness and improved drapeability
- Brighter colors (color revival or disappearance of the grayish look)
- Improved resistance to fuzzing and pilling

The primary objective of **test 3** was to determine how enzymatic bio-finishing of cotton printcloth with various concentrations of cellulase enzyme (1.0 - 8.0 g/L) would be affected by sonication. Triplicate samples of desized, scoured, bleached and mercerized cotton printcloth (107 g/m^2) were treated with cellulase solution (Cellusoft L)™ in combination with mechanical agitation/

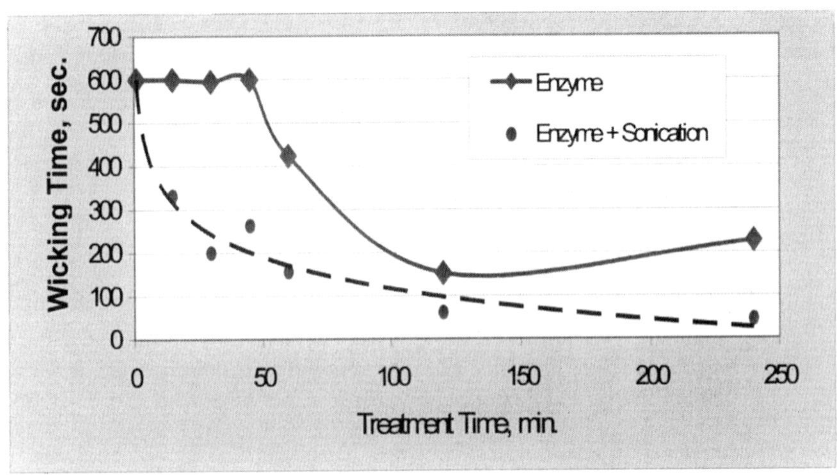

Figure 9. Evaluation of the influence of treatment time on average wettabilities (Warp + Fill)/2 of cotton printcloth samples after alkaline pectinase bio-scouring under sonication conditions (test 2).

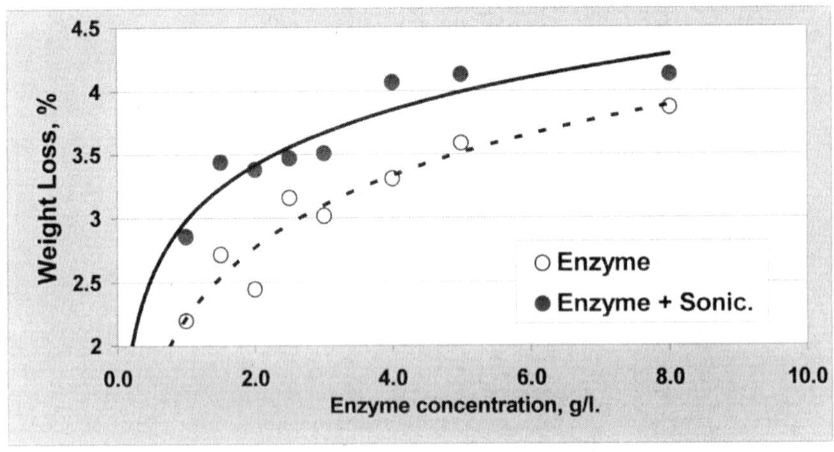

Figure 10. Effect of combined enzyme/ultrasound bio-finishing treatment on weight loss of cotton printcloth (test 3).

circulation and/or sonication for 60 min (Liquor Ratio = 200:1). Results of **test 3** are presented in Figure 10.

For all concentration ranges, the average weight loss of samples treated with a combination of ultrasound and Cellusoft L enzyme was significantly greater then for samples treated with enzyme only. The maximum benefit provided by sonication occurred at concentrations of Cellusoft L of 1-3 g/L, with improvement up to 35% compared to about 20% for enzyme concentrations of 4-8 g/L.

The objective of **test 4** was to determine how enzymatic bio-finishing of different types of cotton fabrics treated with Cellusoft L would be affected by ultrasound. Triplicate samples of mercerized cotton printcloth, combed batiste (70 g/m^2), and mercerized broadcloth (126 g/m^2) were treated with Cellusoft L enzyme with/without sonication. All other treatment parameters in **test 4**, such as sonication power, treatment time, temperature, and circulation rate, were identical to those in **test 3**. The experimental data presented in Figure 11 showed that the average weight loss of the samples of cotton printcloth, batiste, and broadcloth that were sonicated and enzyme treated increased significantly, up to 34.1%, 108.6% and 28.1% respectively. The improvement in performance of Cellusoft L on light type of fabric (batiste) is more pronounced than for heavier fabrics like broadcloth.

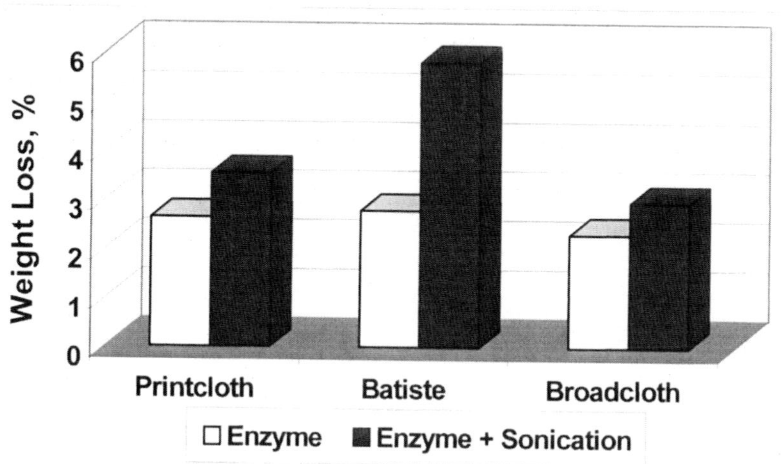

*Figure 11. Effect of combined enzyme/ultrasound bio-finishing treatment on weight loss of cotton printcloth, batiste and broadcloth (**test 4**).*

Discussion

The general trend that was observed in all enzyme bio-preparation and bio-finishing tests indicated that introduction of ultrasound energy during enzymatic bio-treatment of cotton resulted in substantial improvement in the enzyme's performance. The beneficial effects of an introduction of ultrasonic energy in the reaction chamber for enzymatic bio-processing of cotton are:

a) Acceleration of the transport of enzyme macromolecules toward the fiber surface through the border layer of liquid at the liquid-solid interface. The concentration of enzyme macromolecules in this layer is a controlling factor, which defines the overall rate of reaction,

b) Vigorous agitation of the normally immobile border layer of liquid at the liquid-solid interface caused by sonication help enzyme macromolecules more easily position themselves "*fittingly*" onto the substrate,

c) Prevention of possible agglomeration of enzyme macromolecules, that could decrease enzyme activity.

d) Improved removal of the products of enzymatic hydrolysis from the reaction zone, that could accelerate the overall rate of enzymatic reaction,

e) "*Opening up*" of the surface of fibers as result of mechanical impacts produced by collapsing cavitational bubbles,

f) Improved transport of enzyme molecules through the cotton yarns, that results in a more even enzymatic treatment of cellulose fibers, and minimizes the loss of tensile strength of the fabric.

Conclusions

- Sonication of the enzyme processing solution does not reduce the specific activity of enzyme macromolecules significantly.

- On a laboratory scale, introduction of ultrasonic energy in the reaction chamber during enzymatic bio-preparation and bio-finishing of cotton resulted in a significant improvement in enzyme efficiency.

- The combination of enzymatic bio-preparation and bio-finishing with sonication could significantly advance the new "green chemistry" process and make it more suitable for widespread industrial implementation. This could greatly reduce the amount of wastewater effluents, energy consumption, and overall processing costs.

- Ultrasound has good potential to intensify other technological processes that involve various types of enzymes and matching substrates. Practically any solid/liquid system that involves a reaction between enzyme macromolecules and the solid substrate would greatly benefit from introduction of ultrasonic energy into the system.

Disclaimer: Specific company, product, and equipment names are given to provide exact description of experimental details. Their mention does not imply recommendation or endorsement by the U.S. Department of Agriculture.

References

1. Aehle, W. In *Enzymes in Industry*, Aehle, W.; Ed.; Wiley-VCH Verlag GmbH and Co. KGaA: Weinheim, **2004**, p 219-231.
2. Lange, N. K. *Book of Papers, AATCC International Conference and Exhibition,* **1996**, 101-108.
3. Sarkar, K.; Etters, J. N. *AATCC Review*, **2001**, *1*, 48-52.
4. Yonghua, L.; Hardin, I. R. *Textile Chemist and Colorist,* **1997**, *28*, 71-76.
5. Yachmenev, V. G.; Blanchard, E. J.; Lambert, A. H. *Industrial and Engineering Chemistry Research*, **1998**, *37(10)*, 3919-3923.
6. Yachmenev, V. G.; Blanchard, E. J.; Lambert, A. H. *Ultrasonics*, **2004**, *42*, 87-91.
7. Mark, G.; Tauber, A.; Laupert, R.; Schechmann, H.-P.; Schulz, D.; Mues, A.;Von Sonntag, C. *Ultrason. Sonochem.* **1998**, *5*, 41-52.
8. De Gennaro, L.; Cavella, S.; Romano, R.; Masi, P. *J. Food Eng.* **1999**, *39*, 401-407.
9. Lopes, P.; Sala, F.J.; De la Fuente, J. L.; Condon, S.; Raso, J.; Burgos, J. *J. Agric. Food. Chem.* **1994**, *42*, 252-256.
10. Sala, F.J.; Burgos, J.; Condon, S.; Lopes, P.; Raso, J. *Book of Papers, New Methods of Food Preservation.* **1995**, 176-204.
11. Traore, M. K.; Buschle-Diller, G. *Textile Chemist and Colorist and American Dyestuff Reporter*, **1999**, *1*, 51-56.
12. Suslick, K. S. In *Ultrasound: Its Chemical, Physical and Biological Effects*, Sulick, K. S.; Ed.; VCH Publishers: New York, **1988**.

13. Mock, G. N.; Klutz, D. S.; Smith, C. B.; Grady, P. L.; McCall, R. E.; Cato, M. J. *Book of Papers, AATCC International Conference & Exhibition,* **1995**, 55-64.
14. Thakore, K. A.; Smith, C. B.; Clapp, T. G. *American Dyestuff Reporter*, **1990**, *79*, 30-44.
15. Hua, I.; Höchemer, R. H.; Hoffmann, M. R. *Environmental Science & Technology*, **1995**, 29, 2790-2796.
16. Bailey, D. L.; Benge, K. D.; Blanton, W. A.; Bowen, M.; Harrison, T. H.; Strahl, W. A.; Turner, J. D.; Tyndall, R. M. In *Cotton Dyeing and Finishing: A Technical Guide,* Strahl, W. A.; Ed.; Cotton Incorporated: New York, **1996**.
17. Lange, N. K.; Liu, J.; Husain, P.; Condon, B. *Enzyme Business,* **1999**, *10*, 1-12.

Chapter 11

Biomimicking of Glucose Oxidase for Bleaching of Cotton Fabric

X. Ren and G. Buschle-Diller

Department of Polymer and Fiber Engineering, Auburn University, Auburn, AL 36849

Glucose oxidases belong to the enzyme class of oxidoreductases, and are capable of generating hydrogen peroxide *in-situ* with glucose as the substrate. In previous work, enzymatically produced hydrogen peroxide has been used for bleaching of cotton fabric. The active site of glucose oxidase contains flavin adenine dinucleotide (FAD). Compounds similar to FAD such as flavin mononucleotide (FMN), riboflavin, and lumiflavin were used in an attempt to mimic the reactions of the intact enzyme. The effect of amino acids in the vicinity of FAD, as well as the influence of pH and light sources with different energy output, was investigated. Cotton fabric was used as an indicator for whiteness that could be achieved with the mimics. A level of whiteness of approximately 60-70% could be reached with the mimics, compared to the intact enzyme under comparable conditions.

Introduction

Bleaching is one of the preparatory processing steps routinely performed in the textile industry for cotton and other cellulosic fabrics, with the purpose of removing natural pigments and other noncellulosic impurities. A high level of whiteness is essential to guarantee reproducible color shades when the fabric is dyed. Currently, hydrogen peroxide, applied under high pH and boiling conditions, is the most common bleaching agent used. An alternative eco-friendly process to chemical bleaching, is the use of hydrogen peroxide that can be enzymatically produced by glucose oxidase (GO) (*1,3*).

Glucose oxidase (EC1.1.3.4) belongs to the group of oxidoreductases with flavin prosthetic groups. Each subunit of the enzyme contains one mole of flavin adenine dinucleotide (FAD) (Figure 1).

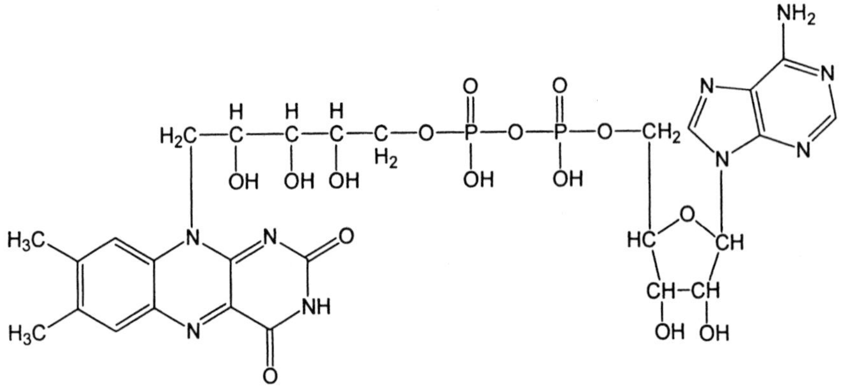

Figure 1. Chemical structure of flavin adenine dinucleotide, FAD.

The enzyme is highly specific for β-D-glucose, and can generate hydrogen peroxide in the presence of oxygen by using β-D-glucose as the substrate (*1-4*) (Figure 2). β-D-glucose is oxidized to gluconic acid with the transfer of two protons and two electrons from the substrate to the flavin moiety. The produced hydrogen peroxide can then act as a bleaching agent for the decomposition of any yellowing compounds in cotton or other cellulosics.

Glucose oxidase is a very large complex protein molecule with a highly specific active site that catalyzes selected redox reactions. The active site, consisting of the prosthetic group FAD and amino acid residues adjacent to FAD, determines the specificity of the glucose oxidase in redox reactions. Flavins can undergo either two sequential one-electron transfers, or a simultaneous two-electron transfer through the semiquinone state. The reaction of GO with glucose can be seen as a reductive half-reaction (hydride transfer from C-H of glucose to FAD) and an oxidative half-reaction in which the reduced FAD ($FADH^-$) is oxidized by O_2 generating H_2O_2. The apoprotein of glucose oxidase most likely plays an important role in the catalytic reaction due to the pH dependency of the reduction potentials (*5-6*). Of the amino acid residues, histidine seems to be most influential on the catalytic reaction of the reduced $FADH^-$ and oxygen (*7-8*).

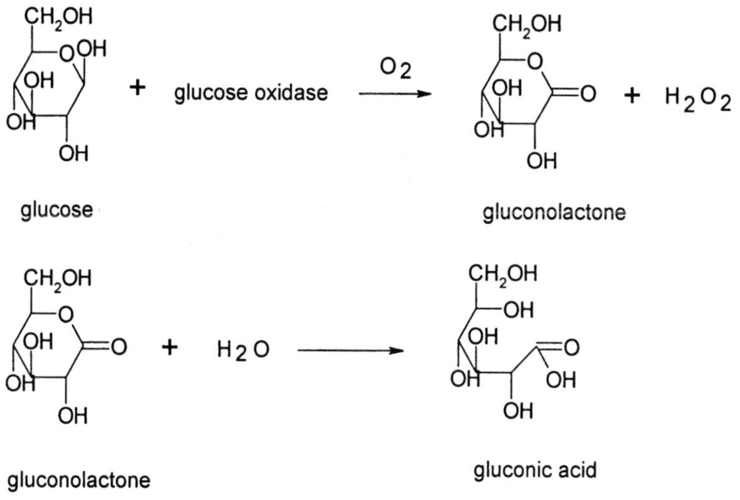

glucose + glucose oxidase $\xrightarrow{O_2}$ gluconolactone + H_2O_2

gluconolactone + H_2O $\longrightarrow$ gluconic acid

Figure 2. Reaction of glucose oxidase and glucose.

Meyer (*9*) reported that the protonated His 516 and His 559 under acidic conditions increases the rate of the reaction between GO and glucose. It has been reported that His 516 is largely responsible for catalyzing the oxidation of $FADH^-$ by oxygen (*8,10*).

It is well-known that vitamin B_2 (riboflavin) plays an important role regarding light sensitivity in biological systems (*11*). Further research showed that hydrogen peroxide could be produced with flavin compounds and, to a lesser extent, methylene blue in the presence of visible light (*12*). Catalase was used to demonstrate that hydrogen peroxide clearly was one product of the system. In earlier studies (*13-14*), it was found that riboflavin, flavin mononucleotide, or lumiflavin could act as photosensitizer, but the addition of compounds such as thiourea, EDTA, or semicarbazide were needed to increase the efficiency of reducing the oxidized isoalloxazine ring. It was assumed that the flavin semiquinone formed in the process transferred the acquired electron to molecular oxygen and to the superoxide anion, to produce hydrogen peroxide.

The mechanism of the light-sensitized reaction has been explained by Heelis *et al.* (*15*) and Fontes *et al.* (*14*) by a one-electron transfer. The flavin semiquinone form produced might move the acquired electron to molecular oxygen and to the superoxide anion, to produce hydrogen peroxide. The photoreduced dihydroflavin (H at N1 and N10 position) can either be oxidized by the presence of oxygen directly or indirectly via a radical route (*16*). A high pH value and the disproportionation of superoxide to hydrogen peroxide might

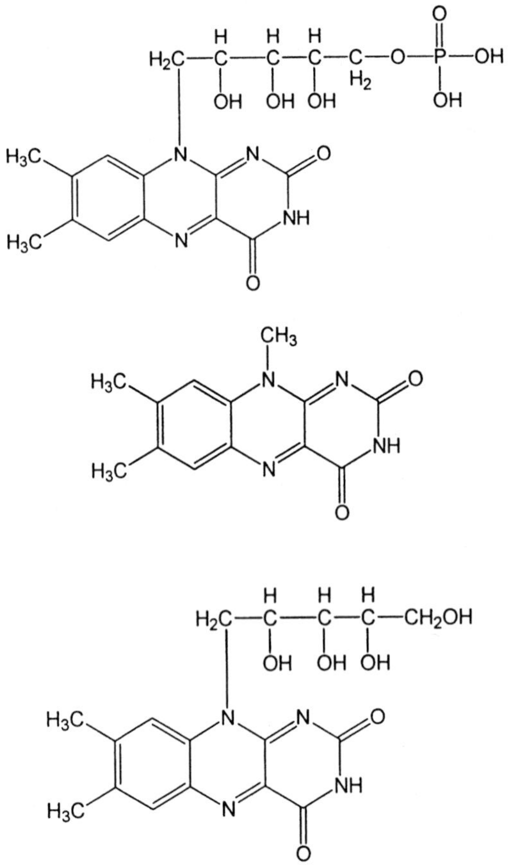

(A)

(B)

(C)

Figure 3. Structure of flavin mononucleotide (A), lumiflavin (B), and riboflavin (C).

favor the reaction via superoxide. The reaction of oxygen with the anionic flavin radical should be faster than with semiquinone in neutral radical form (*17*).

Replacing enzymes with simpler compounds that mimic the behavior of these biocatalysts would help to better understand the mechanism of the enzymatic process. In this book chapter, the active site of glucose oxidase was mimicked by using flavin mononuleotide (FMN), riboflavin, and lumiflavin (Figure 3). Semicarbazide, originally used as electron donor, was replaced by amino acids that are either adjacent to FAD in the original enzyme, or that could assist the reaction of glucose oxidase and glucose. The effect of a light source during the mimicking reaction was also taken into account.

Experimental

Materials

Glucose oxidase (202 U/mg), lumiflavin, riboflavin, flavin mononucleotide (FMN), semicarbazide, L-histidine, L-lysine, L-asparagine, and L-arginine were purchased from Sigma-Aldrich Chemicals. Sodium acetate, sodium hydroxide, glucose, and all other chemicals (reagent grade) were obtained from Fisher Chemicals. The weight per unit area of 100% scoured unbleached cotton fabric (Testfabrics, Inc., New Jersey) was 109.2 g/m^2.

Enzymatic Reactions

A series of samples was produced by using the intact enzyme. The conditions for producing hydrogen peroxide by glucose oxidase (GO) were set to 35 °C and pH 5.1 (0.05 M aqueous sodium acetate buffer). Glucose dosage was 10 g/L and reaction time was adjusted to 2 h. The concentration of glucose oxidase was varied. The concentration of active hydrogen peroxide was determined according to AATCC Test Method 102-1992 by titration with standardized 0.588N potassium permanganate ($KMnO_4$) solution in acidic medium. The bleaching process was performed at pH 10.5 and 90-95 °C for 2 h with the enzymatically produced hydrogen peroxide. The samples were neutralized with dilute acetic acid, washed in water, air-dried and the color coordinates determined using a colorimeter. The control sample for comparison of whiteness levels was produced with 10 U/mL GO.

Reactions with Mimicking Compounds

Aqueous solutions (60 mL) containing 10 mL 4.0 x 10^{-4} M mimicking compounds (lumiflavin, riboflavin, or FMN) were prepared. As electron donors either semicarbazide (SC, 15 mM), or one or more of the amino acids L-histidine (His), L-lysine (Lys), L-asparagine (Asp), and L-arginine (Arg) were added (concentration varied, see text). For experiments with amino acid(s) the treatment time was 8 h and pH 5.1 (buffer obtained from aqueous solutions of 0.20 M boric acid, 0.05 M citric acid solution and 0.10 M tertiary sodium phosphate). The glucose dosage was set to 10 g/L. Oxygen was supplied in gaseous form. The treatment solutions were irradiated with a 75 W white light.

For experiments involving semicarbazide the solutions were adjusted to pH 12.3 or 13.3 with sodium hydroxide and irradiated with a 60 W halogen light or 75 W white light for various lengths of time. Oxygen was supplied the same as mentioned above. The concentration of hydrogen peroxide in the bleach bath was determined by an iodometric method (*18*). For the actual bleaching step, the pH was adjusted to 7 or 10.5, and the reaction performed at 90-95 °C for 2 h with the hydrogen peroxide generated by the mimics. After bleaching, the samples were thoroughly washed in water and air-dried. Their color coordinates were compared with those of the control samples. All experiments were repeated multiple times.

Fabric Whiteness

The level of whiteness of control and treated samples was measured with a color spectrophotometer (CS-5 Chroma Sensor, Datacolor International). Based on the CIEL*a*b* color system, DL* (difference in whiteness) was used to represent the bleaching effect after treatment compared to the control.

Results and Discussion

Bleaching Cotton Fabric with Glucose Oxidase

For commercial hydrogen peroxide bleaching of cotton fabric the amount of hydrogen peroxide added to the bleaching bath is related to the impurities of the fabric and the whiteness level required. The pH value of the bleach bath generally ranges from 10 to 11 and the temperature from 90 to 100 °C. These conditions are fairly harsh and fiber damage could occur, especially in the presence of metal ions introduced by rusty equipment.

For glucose oxidase bleaching, in order to achieve the desired whiteness, sufficient hydrogen peroxide must be produced by GO and glucose under aerobic conditions. Both the amount of glucose and GO dosage play an important role in producing hydrogen peroxide. In previous work, glucose dosage was determined to be sufficient at 10 g/L (*1*). Figure 4 shows the relationship between GO dosage and hydrogen peroxide concentration with/without fabric present during the reaction. By increasing the GO dosage from 2 to 10 U/mL higher concentrations of peroxide could be achieved. Above 10 U/mL GO the rate of peroxide production decreased as previous research had demonstrated (*1*). It is interesting that the presence of fabric during this step played a considerable role in producing peroxide. Without the fabric the amount of peroxide produced was between 800 and 950 mg/L while with fabric 1120-1260 mg/L peroxide was generated.

The pH of the treatment bath generally is crucial for all enzymes to perform at their optimum. In the case of the interaction of GO with glucose, slightly acidic conditions (pH 5.1) are optimal. Hydrogen peroxide for bleaching, on the other hand, requires for the pH to be adjusted to 7.0 (*1*) or 10.5 (industrial conditions) to be effective. In Figure 5 the relationship between whiteness of enzymatically bleached fabric, GO dosage, and pH setting is shown. The fabric whiteness was higher at increased GO concentration due to more available

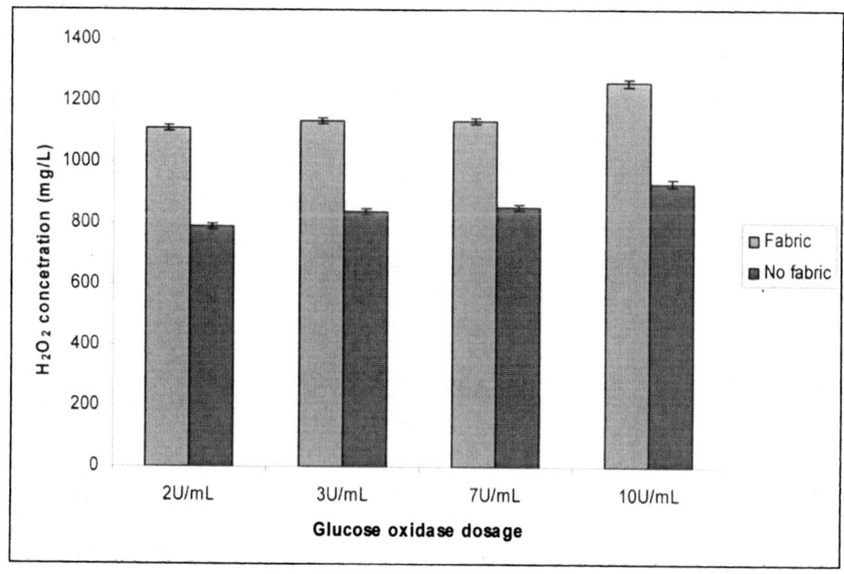

Figure 4. Enzymatic production of hydrogen peroxide with and without fabric with increasing dosage of GO.
(See page 5 of color inserts.)

hydrogen peroxide produced at higher GO dosage. The bleaching effect of the bath at pH 10.5 was enhanced compared to pH 7 as could be expected.

Light-mediated Biomimicking of FAD with an Electron Donor

FAD, the active site of GO, and amino acid residues adjacent to FAD, catalyze the reaction with glucose in the process of which hydrogen peroxide is generated. The structures of FMN, riboflavin, and lumiflavin are similar to FAD. Thus, these compounds were selected to mimic the behavior of FAD in reaction. De la Rosa *et al.* (*13-14*) performed studies concerning light-sensitized hydrogen peroxide production by flavin compounds and explored the effect of electron donors for the reaction. In their work, semicarbazide had proved to be most efficient. Thus, for this work semicarbizide was initially selected as electron-donor. Additionally, a light source with defined energy output was installed. It had also been established (*13*) that the photochemical formation of hydrogen peroxide is greatly influenced by the pH value. Higher pH values favor the rate of hydrogen peroxide production.

In Figure 6, the amount of photochemically produced hydrogen peroxide by riboflavin is presented in relation to the irradiation time at pH 12.3 and 13.3 (white lamp 75 W). At pH 12.3 the concentration of hydrogen peroxide

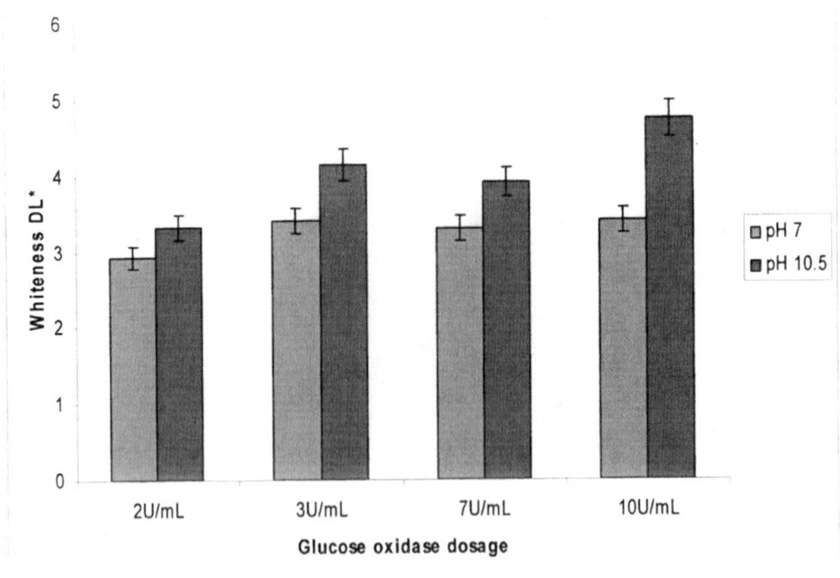

Figure 5. Whiteness increase of cotton fabric based on GO concentration compared to the unbleached control; at two pH settings (7, 10.5). The fabric was present in the bath during the entire process.
(See page 5 of color inserts.)

increased during the first four hours of the reaction, then leveled off. Hydrogen peroxide production at pH 13.3 was almost three times higher than at pH 12.3 and kept increasing with reaction time.

The photochemically produced hydrogen peroxide was used to bleach cotton fabric after lowering the pH. In Figure 7, results for fabric whiteness are presented obtained with riboflavin at pH 10.5 and 7. At pH 7 a steady increase of fabric whiteness with time was observed, while at pH 10.5 whiteness reached a maximum at 8 h corresponding to the amounts of available hydrogen peroxide. However, even with such extended treatment times, only approximately 70% of the whiteness levels of the enzymatically (GO) treated control could be achieved.

The efficiency of the photochemical system might be related to the type of mimicking compound as well as the light source with its specific energy output (Table I). Compared to riboflavin and lumiflavin, FMN was more effective in whitening cotton fabric with the 60 W halogen light source, even with the concentrations of hydrogen peroxide being slightly lower, and lumiflavin performed best with the 75 W white light. Overall, the results of both light sources were similar.

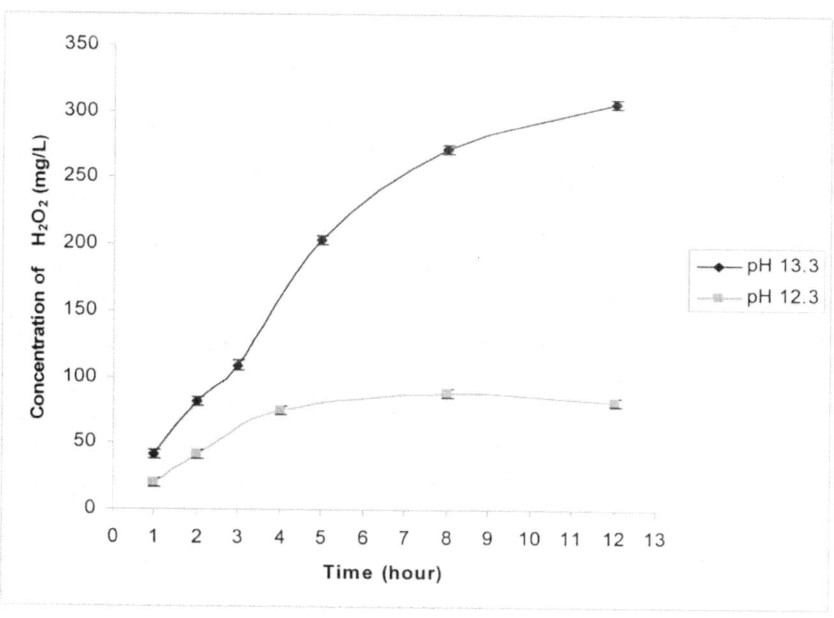

Figure 6. Light-sensitized production of hydrogen peroxide by riboflavin with reaction time at pH 12.3 and 13.3 (white lamp 75 W).
(See page 6 of color inserts.)

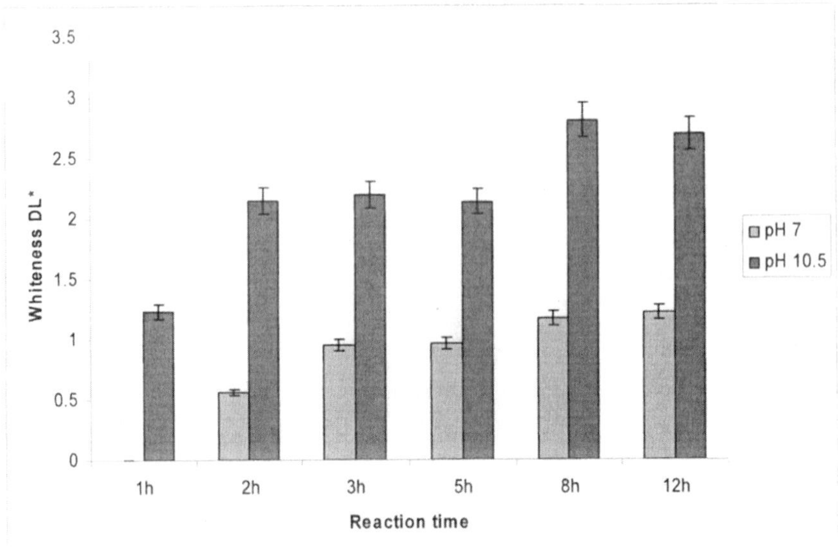

Figure 7. Whiteness increase of cotton fabric treated with riboflavin as a function of treatment time and pH compared to the scoured control (pH value of previous step: 13.3).
(See page 6 of color inserts.)

Table I. Production of hydrogen peroxide and whiteness increase of cotton fabric with different mimics irradiated by 75 W white and 60 W halogen light sources (irradiation 8 h).

Light source		*Riboflavin*	*Lumiflavin*	*FMN*
White lamp	H_2O_2 (mg/L)	272 ± 3.4	286 ± 3.4	258 ± 3.4
75 W	DL*	2.81 ± 0.14	2.93 ± 0.15	2.77 ± 0.13
Halogen	H_2O_2 (mg/L)	272 ± 3.4	258 ± 3.4	258 ± 3.4
60 W	DL*	2.47 ± 0.12	2.82 ± 0.15	3.07 ± 0.15

Biomimicking of FAD with Amino Acids as Electron Donors

Amino acid residues in direct vicinity of FAD are believed to play a vital role in the reaction as electron donors. For this research, histidine (His), lysine (Lys), aspartic acid (Asp), and arginine (Arg) were used to mimic the environment of FAD and to support the electron transfer. The pH was originally varied in the range of 4 to 10, including pH 5.1 which is the optimum pH for the intact GO used in the control experiments. Higher or lower pH values resulted in lower performance of the mimics; thus, only results at pH 5.1 are reported here. The energy output of the light source was also varied, but only results obtained with a 75 W white light for irradiation are presented in this paper since the difference between light sources was insignificant.

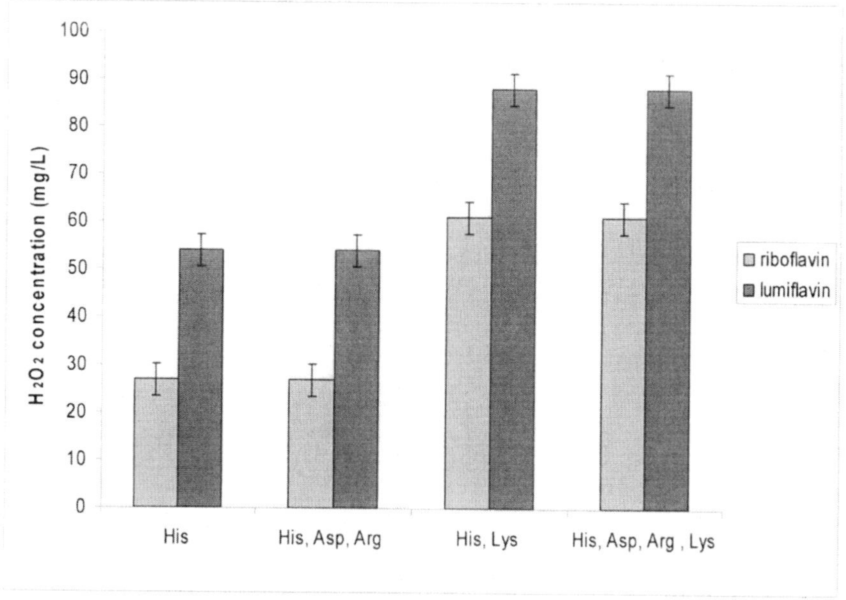

Figure 8. Production of hydrogen peroxide by riboflavin and lumiflavin with various combinations of His, Lys, Asp, and Arg in equal mole ratio. The amount of each amino acid used in this experiment was 3.2 x 10^{-4}M. *(See page 7 of color inserts.)*

In Figure 8, the effect of hydrogen peroxide concentration with regard to different combinations of these amino acids and mimics is presented. Equal molar ratios of each amino acid in combination were used. The results indicate that His and Lys were more effective than His alone or the combination of His,

Asp, and Arg in producing peroxide. The addition of Arg and Asp to His and Lys did not significantly improve hydrogen peroxide production.

Regarding the mimics, lumiflavin seemed to be more effective than riboflavin. In Figure 9, the relationship of hydrogen peroxide concentration and the different flavin mimics under otherwise same conditions are compared.

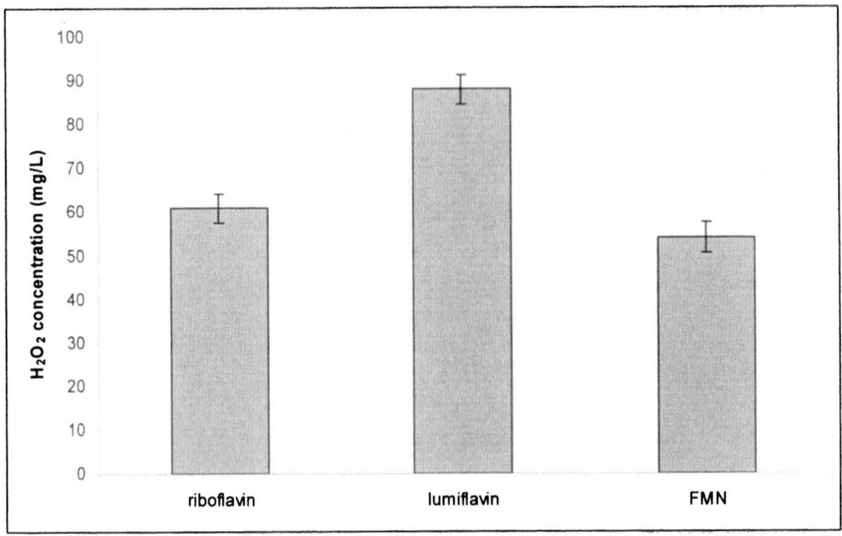

Figure 9. Production of hydrogen peroxide by lumiflavin, riboflavin, and FMN with a mole ratio of His and Lys of 1:1.
(See page 7 of color inserts.)

Since the combination of His and Lys seemed to work best to mimic the environment of FAD, and His 516 and His 559 are in the direct vicinity of FAD in the intact GO (*8*), different molar ratios of His and Lys were studied further.

Table II. Production of hydrogen peroxide by lumiflavin with varied dosage of His and Lys (mole ratio 2:1).

His (mole)	*Lys (mole)*	*H_2O_2 (mg/L)*
2.0×10^{-4}	1.0×10^{-4}	40 ± 3.4
5.0×10^{-4}	2.5×10^{-4}	102 ± 3.4
6.4×10^{-4}	3.2×10^{-4}	107 ± 1.7
9.0×10^{-4}	4.5×10^{-4}	122 ± 1.7
1.8×10^{-3}	9.0×10^{-4}	150 ± 3.4

In Table II the amounts of hydrogen peroxide are listed that were generated in presence of lumiflavin with increasing dosage of His and Lys at a 2:1 mole ratio. Hydrogen peroxide concentration produced by this mimic system increased with the increase of amino acid dosage.

Even though hydrogen peroxide could be produced by the flavin mimics/amino acid system, the relative amount was lower than the intact enzyme could generate (see Figure 4). It was assumed that the electron donor capacity of the free amino acids under the chosen experimental conditions was not quite sufficient. In this aspect, semicarbazide was more effective. The hydrogen peroxide production however occurred at more benign conditions (pH 5.1) with amino acids while semicarbazide only functioned well at highly alkaline pH values.

Conclusions

Hydrogen peroxide has been produced by the reaction of intact glucose oxidase and glucose, and was used for bleaching of scoured cotton. Fabric whiteness levels were comparable to those achieved with commercial non-enzymatic processes. In an attempt to simplify the enzymatic process, the GO/glucose system was mimicked by compounds similar in structure to FAD of the enzyme's active site, such as flavin mononucleotide (FMN), riboflavin, and lumiflavin. These mimics were used for photosensitized reactions in dependence of the pH and light source. Amino acids as electron-donors enhanced the production of hydrogen peroxide within the biomimicking system, with histidine and lysine being the most effective. Semicarbazide previously used as electron donor appeared to be more effective than the amino acids; however, the reactions had to be performed at very high pH values. Overall, bleaching efficiency achieved with the mimics was approximately 60-70% compared to fabric whiteness reached with the integral enzyme under comparable conditions.

References

1. Buschle-Diller, G.; Yang, X.D.; Yamamoto, R. *Text. Res. J.* **2001**, *71(5)*, 388-394.
2. Tzanov, T.; Calafell, M.; Guebitz, G.M.; Cavaco-Paulo, A. *Enz. Micro. Techn.* **2001**, *29*, 357-362.
3. Tzanov, T.; Costa, S.A.; Guebitz, G. M.; Cavaco-Paulo, A. *J. Biotechnol.* **2002**, *93*, 87-94.
4. Haouz, A.; Twist, C.; Zentz, C.; Tauc, P.; Alpert, B. *Eur. Biophys. J.* **1998**, *27*, 19-25.

5. Keene, R.F.; Salmon, D.J.; Meyer, T.J. *J. Amer. Chem. Soc.* **1977**, *99(7)*, 2387-2389.
6. Hille, R.; Anderson, R.F. *J. Biol. Chem.* **2001**, *276(33)*, 31193-31201.
7. Wong, D.W.S. In: *Food Enzymes: Structure and Mechanism*, Wong, D. W. S.; Ed.; Chapt. 10, Chapman and Hall: New York; **1995**, p 308-320.
8. Roth, J.P.; Klinman, J.P. *Proc. Nat. Acad. Sci.* **2003**, *100*, 62-67.
9. Meyer M.; Wohlfahrt, G.; Knablein, J.; Schomburg, D. *J. Comp.-Aided Molec. Design* **1998**, *12(5)*, 425-440.
10. Prabhakar, R.; Siegbahn, P.E.M.; Minaev, B.F. *Biochim. Biophys. Acta* **2003**, *1647*, 173-178.
11. Garcia, J.; Silva, E. *J. Nutr. Biochem.* **1997**, *8*, 341-345.
12. Silva, E.; Edwards, A.M.; Pacheco, D. *J. Nutr. Biochem.* **1999**, *10*, 181-185.
13. De la Rosa, M.A.; Navarro, J.A.; De la Rosa, F.F.; Losada, M. *Photobiochem. Photobiophys.* **1983**, *5*, 93-103.
14. Fontes, A.G.; De la Rosa, F.F.; Gomez-Moreno, C. *Photobiochem. Photobiophys.* **1981**, *2*, 355-364.
15. Heelis, P.F.; Parsons, B.J.; Phillips, G.O.; McKellar, J.F. *Photochem. Photobiol.* **1979**, *30*, 343-347.
16. Singer, T.P.; Edmonson, D.E. In *Molecular Oxygen in Biology*; Hayaishi, O.; Ed.; North-Holland Publ. Company: Amsterdam, **1974**, p 325-330.
17. Draper, R.D.; Ingraham, L.L. *Arch. Biochem. Biophys.* **1968**, *125*, 802.
18. Bassett, J.; Denney, R.C.; Jeffery, G.H.; Mendham, J.; Eds. In *Textbook of Quantitative Inorganic Analysis*, Longman: New York, **1978**, p 381-382.

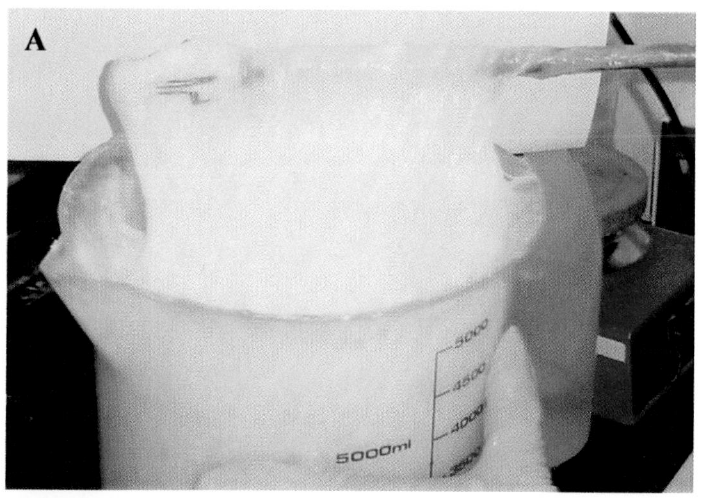

Figure 4.2. Sacchrification of maize amylopectin by a-1,4-glucan layse to 1,5-anhydrofructose (AF). (A). Before dosing of the glucan layse. (B). 3 days after the dosing of the algal GLq1-glucan lyase.

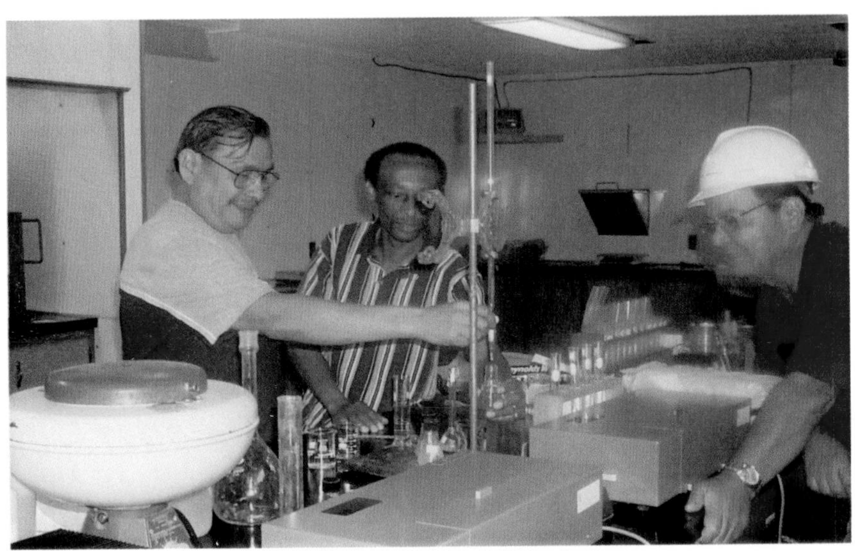

Figure 6.3. Photograph of the use of the Eggleston (14) titration method to measure the activity of commercial dextranases at the factory

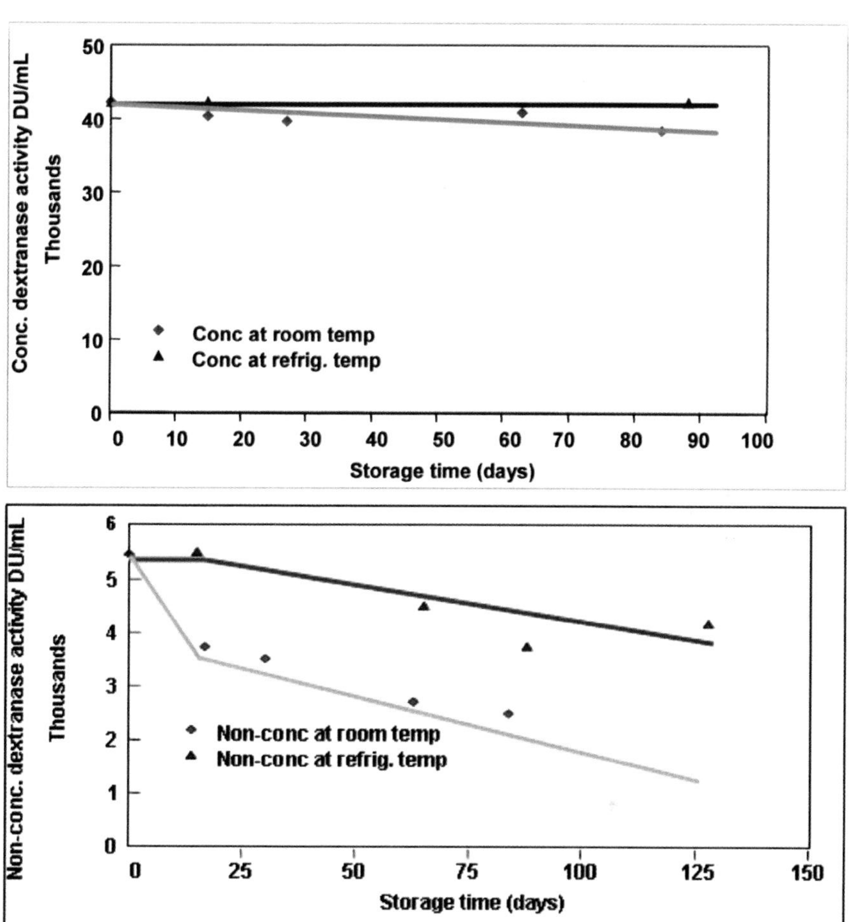

Figure 6.4. Changes in activity of "concentrated" and "non-concentrated" dextranases stored under (top) simulated factory storage conditions (ambient temperature ~25 °C), and (bottom) refrigerated conditions (4 °C) over a 90 day sugarcane processing season. From Eggleston and Monge (1).

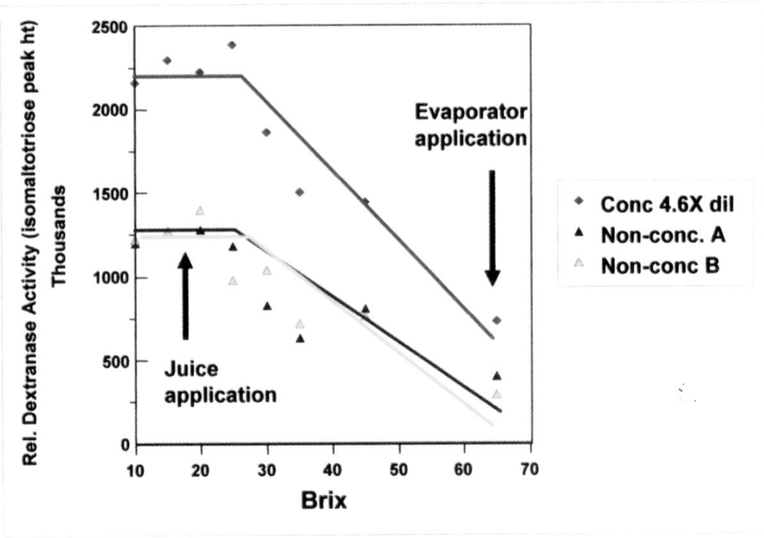

Figure 6.5. Effect of Brix on dextranase activity. The "concentrated" dextranase was diluted 4.6X to make it economically equivalent to the nearest priced "non-concentrated" dextranase. From Eggleston and Monge (1).

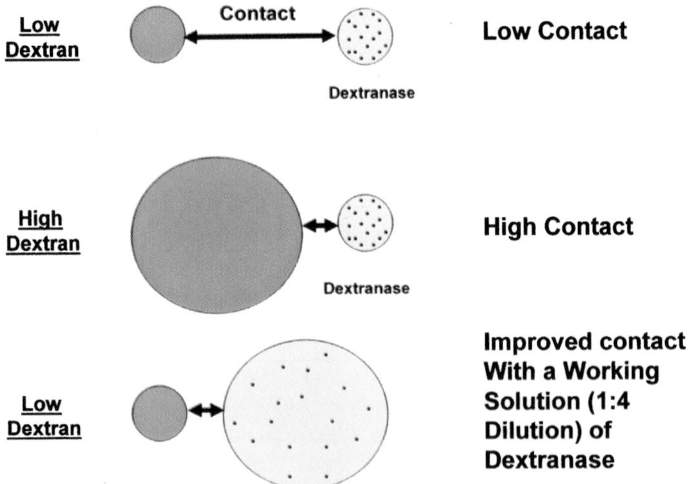

Figure 6.6. Diagram to illustrate the contact between dextranase and different concentrations of dextran. Circles depict volumes and squares depict enzyme molecules. The action of a working solution of "concentrated" dextranase (>25,000 – 58,000 DU/mL) to improve contact in factory process is also shown. Modified from Eggleston et al. (15).

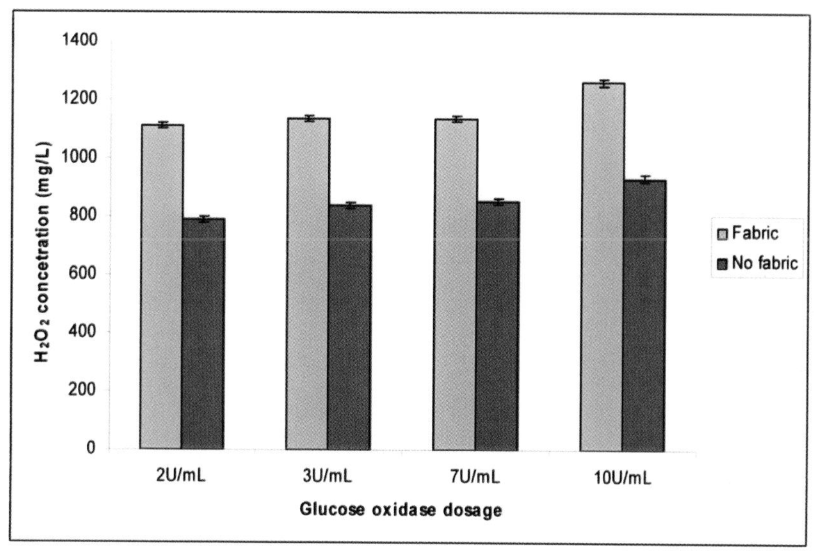

Figure 11.4. Enzymatic production of hydrogen peroxide with and without fabric with increasing dosage of GO.

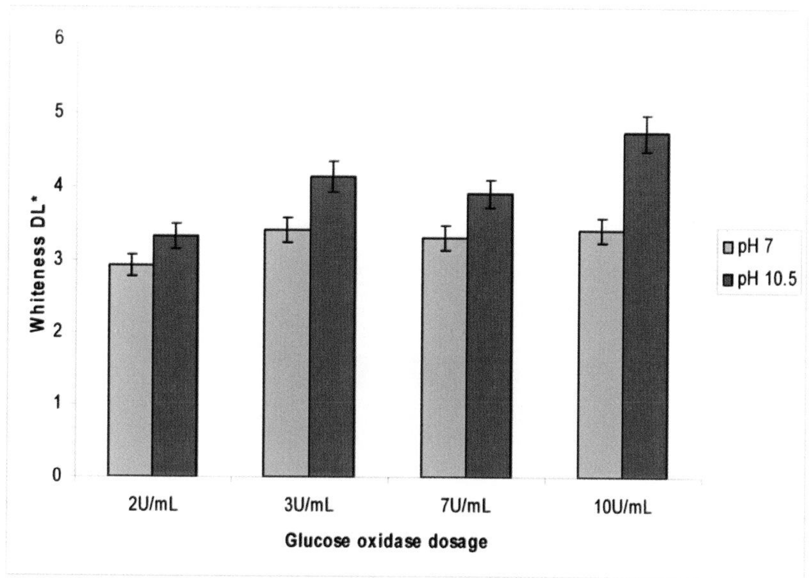

Figure 11.5. Whiteness increase of cotton fabric based on GO concentration compared to the unbleached control; at two pH settings (7, 10.5). The fabric was present in the bath during the entire process.

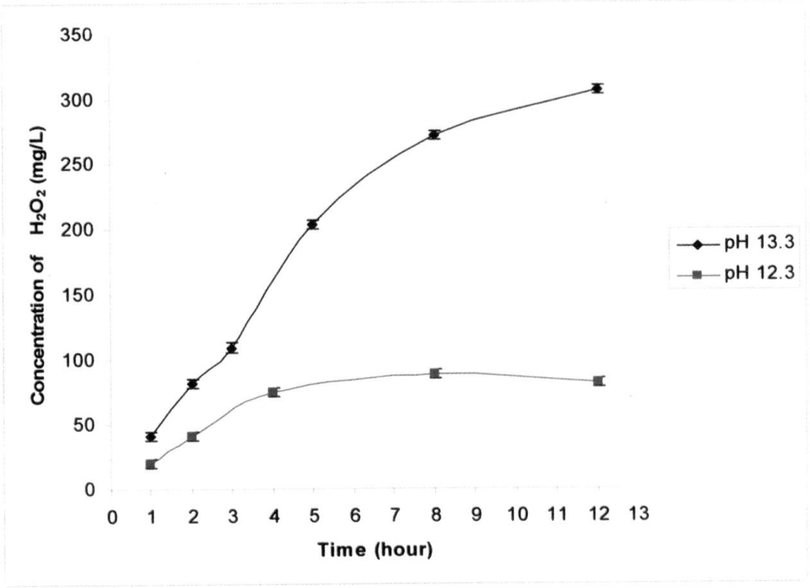

Figure 11.6. Light-sensitized production of hydrogen peroxide by riboflavin with reaction time at pH 12.3 and 13.3 (white lamp 75 W).

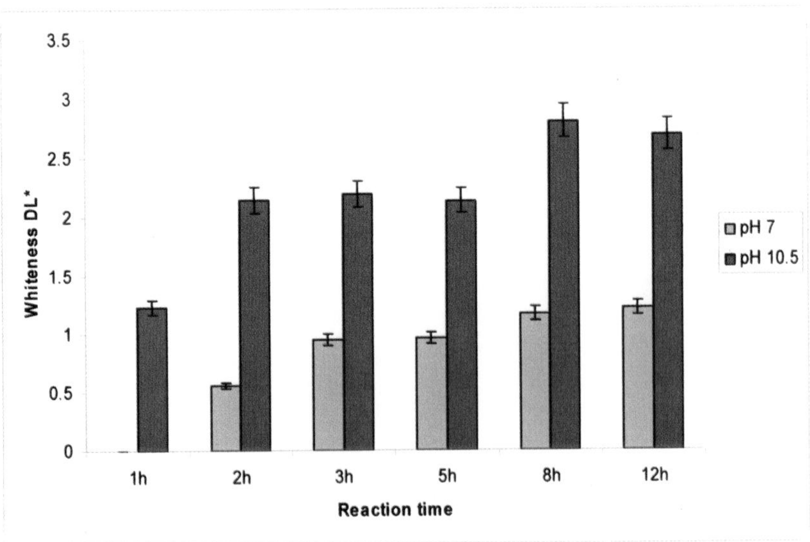

Figure 11.7. Whiteness increase of cotton fabric treated with riboflavin as a function of treatment time and pH compared to the scoured control (pH value of previous step: 13.3).

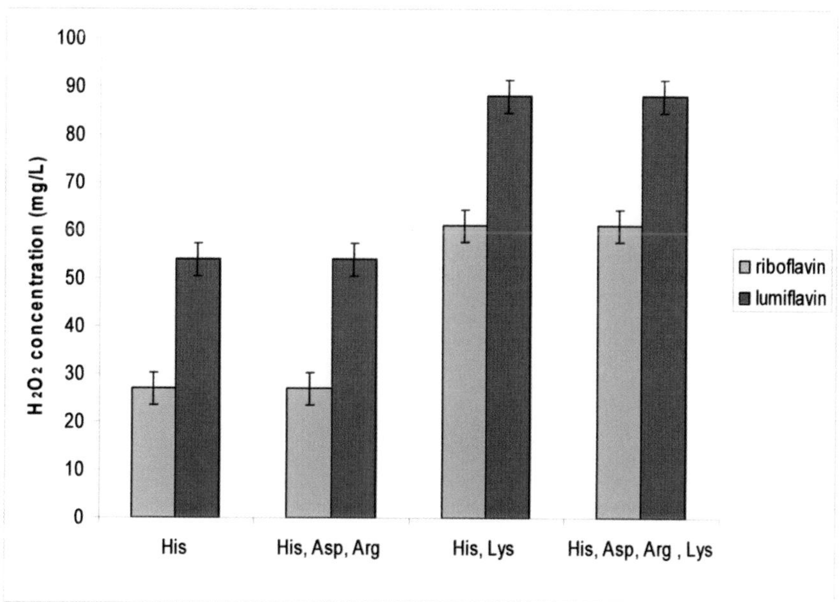

Figure 11.8. Production of hydrogen peroxide by riboflavin and lumiflavin with various combinations of His, Lys, Asp, and Arg in equal mole ratio. The amount of each amino acid used in this experiment was 3.2×10^{-4}M.

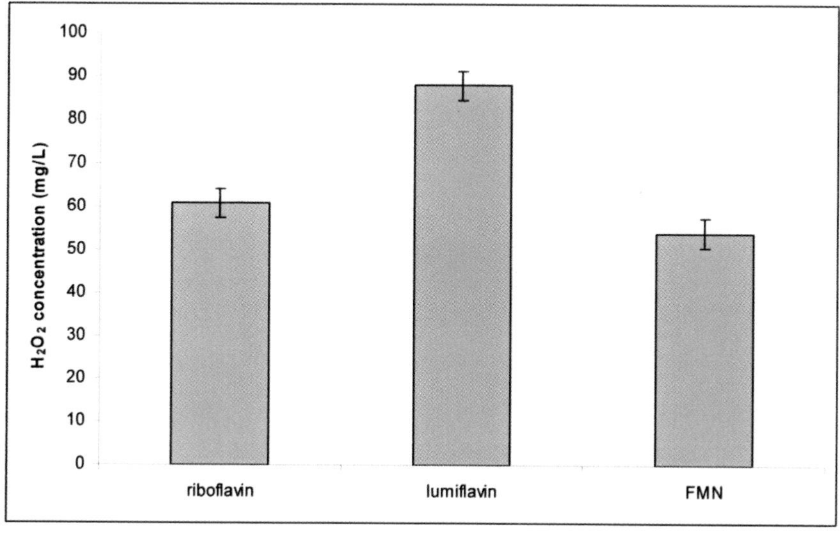

Figure 11.9. Production of hydrogen peroxide by lumiflavin, riboflavin, and FMN with a mole ratio of His and Lys of 1:1.

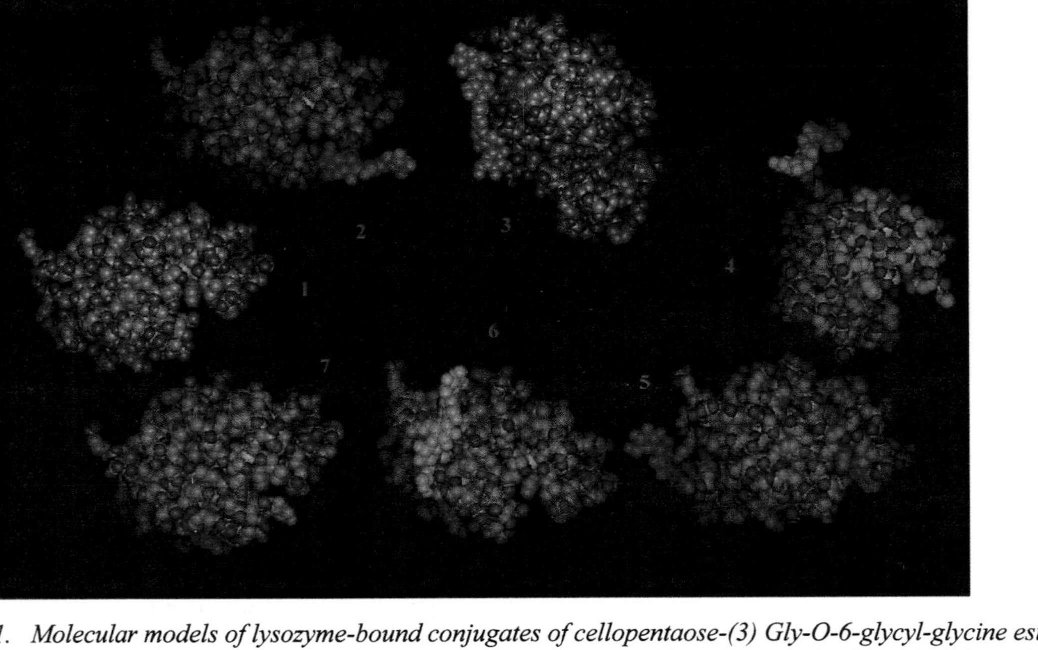

Figure 12.1. Molecular models of lysozyme-bound conjugates of cellopentaose-(3) Gly-O-6-glycyl-glycine ester. The eight structures shown are CPK models. Atom color types are as follows: oxygen (red); carbon (green); nitrogen (blue); sulfur (yellow) and the cellopentaose is highlighted in blue: all models are oriented with the enzyme active site cleft in the right lower quadrant of the protein structure. The lysozyme (1) crystal structure may be compared with glycoprotein conjugates structures 2-7 to visualize potential contact regions between the cellulose (represented here by the cellopentaose) and the lysozyme. Amide linkages to the protein with the glycopeptide are the following: **1. Lysozyme, 2. Asp-48, 3. Asp-66, 4. Asp-87, 5. Asp-119, 6. COOH-129, 7. Glu-35**

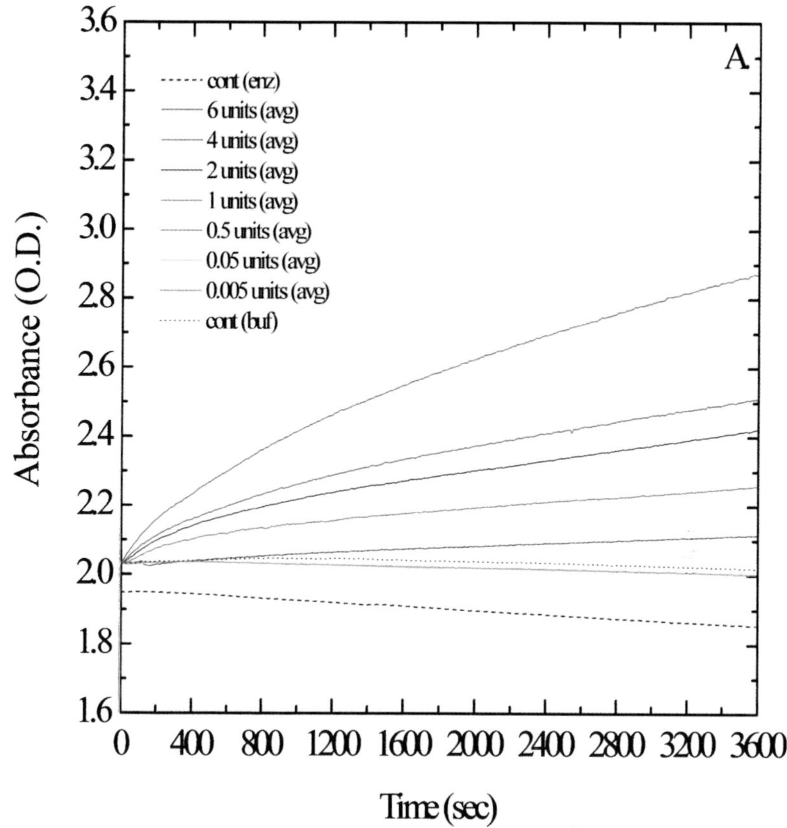

Figure 12.7. Reaction progress curves of the paranitroanilide peptide conjugate Cellulose-APS-suc-Ala-Ala-Pro-Val-pNA to paranitroaniline was monitored at 405 nm (see figure 6 for reaction) for 1 hr. Hydrolysis of substrate was monitored by combining ~ 5mg of sample with the indicated concentration of elastase. The elastase (Athens) was prepared in a phosphate buffer with a pH of 7.6 (0.2 M sodium phosphate, 0.5 M NaCl, and 6.6% DMSO). HNE units of activity employed in the assay ranged from 5.0×10^{-3} units to 6.0 units. In each well of a 96-well microtiter plate 50 μL of enzyme was reacted with ~5 mg of HNE substrate bound paper. The enzymatic hydrolysis of the immobilized substrate on paper was monitored at 405 nm with incubation at 37 °C using a Bio-Rad Microplate Reader for 1 h. The color that developed on the HNE substrate-bound paper was analyzed by Datacolor International Spectraflash 500. Data was obtained in reflectance mode and converted to absorbance.

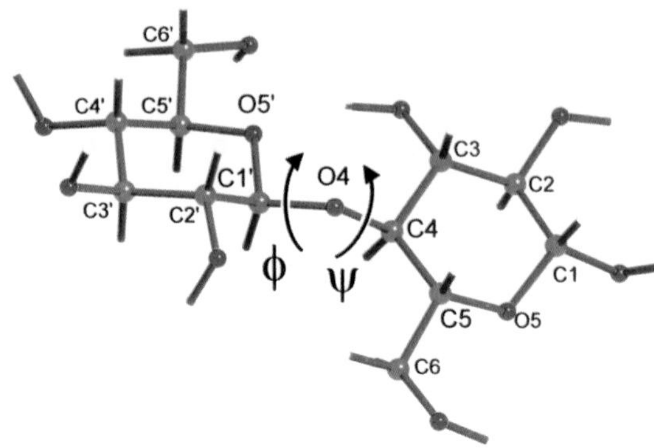

Figure 15.1. β-Cellobiose from its crystal structure (11) showing the linkage torsion angles φ and ψ, as well as the numbering of carbon and ring and linkage oxygen atoms. We define φ and ψ (here φO5'-C1'-O4-C4 = –76.3° and ψC5-C4-O4-C1' = –132.3°) based on the ring atoms O5' and C5 instead of the commonly used H1' and H4 atoms because protons are not found experimentally in most studies of proteins, and are not well located even in many studies of small molecule crystals.

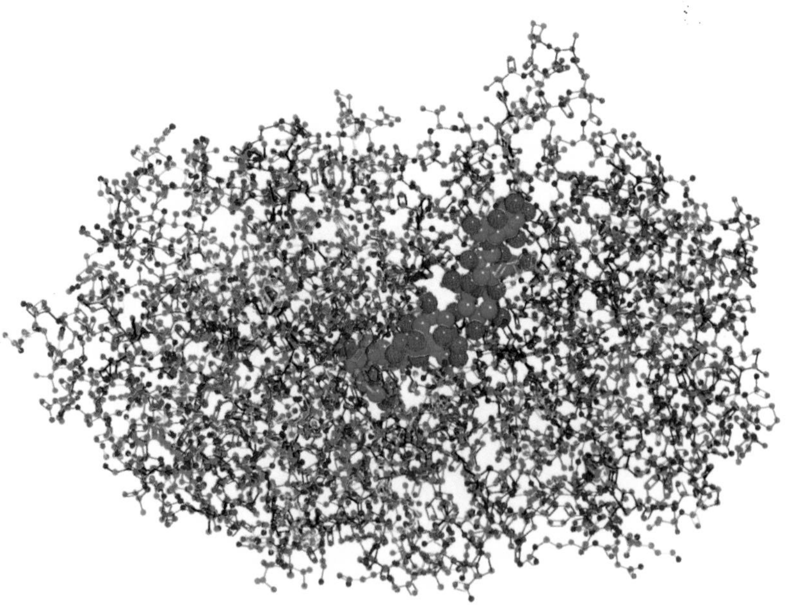

Figure 15.2. The endocellulase E1 catalytic domain (balls and sticks) with a complexed cellotetraose molecule (space-filling model) from the 1ECE crystal structure (12). The 1ECE designation is an ID code from the Protein Data Bank. Each entry has its own ID code. See (13).

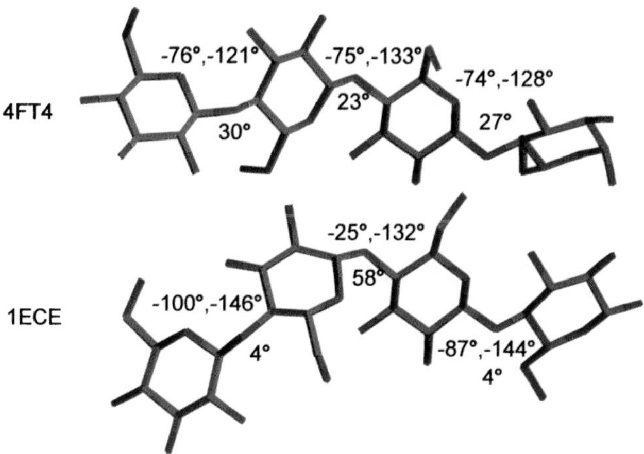

Figure 15.3. Comparison of cellotetraose fragments from the Protein Data Bank. *(See page 211 for the full caption.)*

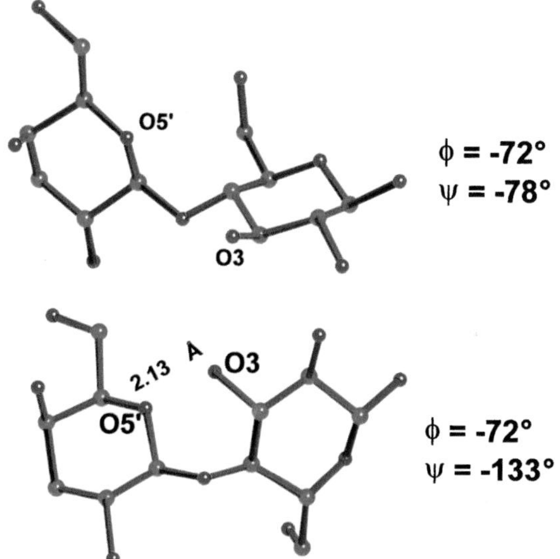

Figure 15.7. Cellobiose-like 3'-deoxy-β-lactose in the reported conformation from the 1FV3 protein complex (upper drawing). See also Figure 5 for the location in φ,ψ space for this linkage. The lack of the 3' hydroxyl group is not expected to substantially influence the linkage conformation. The lower drawing shows the same molecule after rotating the ψ torsion angle from -78° to -133°. There, the distance between O3 and O5' is only 2.13 Å. Since the same distance in the accurately determined cellobiose crystal is 2.77Å, the actual details of the 1FV3 are probably not well determined.

Chapter 12

New Uses for Immobilized Enzymes and Substrates on Cotton and Cellulose Fibers

J. Vincent Edwards, Abul J. Ullah, Kandan Sethumadhavan, Sarah Batiste, Patricia Bel-Berger, Terri Von Hoven, Wilton R. Goynes, Brian Condon, and Sonya Caston-Pierre

Cotton Chemistry and Utilization Research Unit, Southern Regional Research Center, Agricultural Research Service, U.S. Department of Agriculture, New Orleans, LA 70124

The design, preparation, and application of both immobilized enzymes and enzyme substrates on cotton fibers for biomedical and specialty applications, includes antibacterial fabrics, decontamination wipes, debridement and chronic wound dressing prototypes, and protease detection devices. The molecular design steps of enzyme and enzyme substrate cellulose conjugates is presented. Molecular models of lysozyme- and orgaonphosphorous hydrolase- cellulose conjugates are given as examples of assessing the utility of the biorationally designed fabrics. The chemistry of immobilzing the enzyme or enzyme substrate to the cotton fabric of choice, employs use of crosslinking agents for either aqueous or organic coupling reactions. Assays for assessment bioconjugate activity is pivotal to optimizing the development of the immobilized enzyme or substrate as a product.

Enzymes Immobilized on Cotton

The commercial development of immobilized enzymes on cotton has only been realized for a few product applications. For example, the use of β-galactosidase immobilized on individual cotton fibrils of cotton cloth to produce galacto-oligosaccharides from lactose, resulted in higher reactor productivity of free sugars (*1*). A variety of enzyme immobilization techniques including covalent binding, cross-linking, and adsorption have been utilized to adhere enzymes to cotton and explore potential applications. Techniques for the adhesion of whole cells containing invertase and glucose oxidase on cotton cloth and threads have been employed in the food and textile industries (*2-8*). These applications deal mostly with utilizing cotton as a vehicle to provide a more stable and recyclable enzyme. Cotton is a well characterized durable fiber that has been used for medical and hygienic purposes throughout the history of civilization, and its highly absorbent and non-toxic properties make it attractive as a carrier for biologically active fibers. The potential utility of immobilizing numerous enzymes on cotton fabrics as a source of biologically active textile surfaces would offer highly selective functional properties while providing an environmentally acceptable textile finish. Here we consider the immobilization of enzymes on cotton as a vehicle for medical, hygienic, and decontamination applications.

Lysozyme on Cotton as an Antibacterial Agent

Lysozyme has been immobilized on numerous solid supports (*9-14*). The immobilization of lysozyme on cotton as an antibacterial agent has been explored from a molecular design point of view. A molecular model of the attachment of lysozyme to cotton cellulose and consideration of the relevant structure/function implications of surface protein and fiber contact points is shown in Figure 1. The design and preparation of lysozyme on cotton cellulose has led to robust antibacterial fabrics. C-type lysozyme is a 1,4-α-*N*-acetylmuramidase that derives its antibacterial activity from the selective lysis of the cell wall peptidoglycan at the glycosidic bondbetween the C-1 atom of *N*-acetylmuramic acid (NAM) and O-4 of *N*-acetylglucosamine (NAG) (*15-17*). The mechanism of action of lysozyme (Figure 2) has been shown to occur through binding of tri-NAG in a deep crevice or cleft (see Figure 1), which contains the active site of the enzyme, and subsequent carbonium ion-mediated hydrolysis of the glycosidic bond between the NAM and NAG residues (*18*). The point of molecular hydrolysis is shown in Figure 2. The mechanism of action of lysozyme is probably the best studied of any enzyme reported to date. The lytic mechanism of action of hydrolases like lysozyme occurs in the crevice that characterizes the conformation of the protein. This crevice motif is visible in the computational

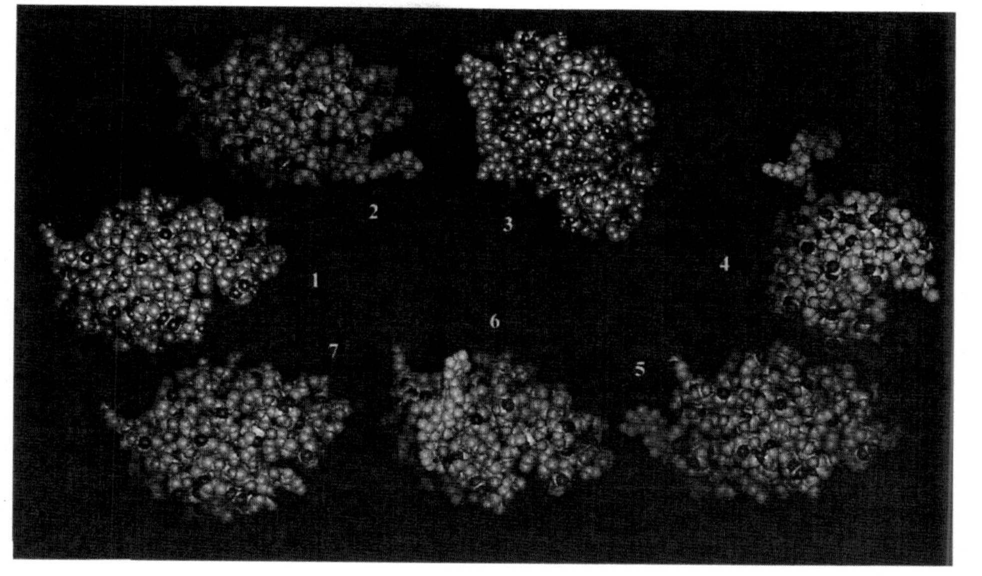

*Figure 1. Molecular models of lysozyme-bound conjugates of cellopentaose-(3) Gly-O-6-glycyl-glycine ester. The eight structures shown are CPK models. Atom color types are as follows: oxygen (red); carbon (green); nitrogen (blue); sulfur (yellow) and the cellopentaose is highlighted in blue: all models are oriented with the enzyme active site cleft in the right lower quadrant of the protein structure. The lysozyme (**1**) crystal structure may be compared with glycoprotein conjugates structures 2-7 to visualize potential contact regions between the cellulose (represented here by the cellopentaose) and the lysozyme. Amide linkages to the protein with the glycopeptide are the following: **1. Lysozyme, 2. Asp-48, 3. Asp-66, 4. Asp-87, 5. Asp-119, 6. COOH-129**, 7. **Glu-35***

(See page 8 of color inserts.)

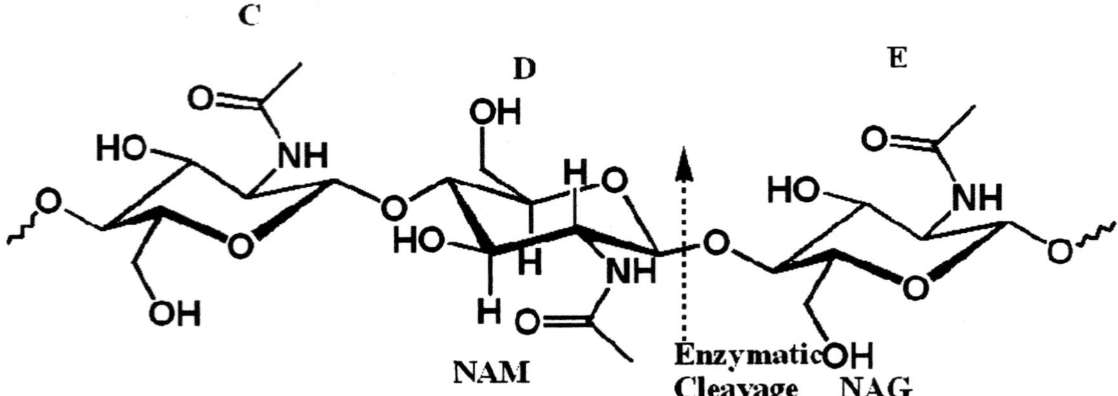

Figure 2. C-type lysozyme is a 1,4-â-N-acetylmuramidase that derives its antibacterial activity from the selective lysis of the cell wall peptidoglycan at the glycosidic bond between the C-1 atom of N-acetylmuramic acid (NAM)and O-4 of N-acetylglucosamine (NAG). The mechanism of action of lysozyme has been shown to occur through binding of tri-NAG in a cleft at the surface of the enzyme and subsequent carbonium ion-mediated hydrolysis of the glycosidic bond between the NAM and NAG residues.

models of Figure 1. Cellulose conjugates that appear to block access to the open crevice of the protein would probably be inactive. Seven lysozyme conjugates of cellopentaose are shown in Figure 1 as illustrations of the general protein and cellulose contact areas, that may occur as a result of covalent attachment of the protein through carbodiimide coupling of glycine-cotton with the carboxylate-containing amino acid side chains of the lysozyme. We have reported that the antimicrobial activity observed on the lysozyme-bound cotton fibers was greater than that observed for lysozyme in solution (*19*). This may be due to the enhanced stability of the enzyme afforded by conjugation to cellulose. Improved stability against proteolytic digestion of proteases released during the course of the bacterial assay may account for this enhanced activity. This might be expected since anchoring of a protein to a solid-phase support can confer protection against proteolytic lysis of the protein.

Enzymes on Fabrics That Come Into Contact with the Skin

Decontamination Towelletes

The conjugation of enzymes to absorptive cotton fabrics that may be applied to human skin for medical or hygienic purposes has received growing interest in recent years. Textiles including towelettes, gauze, swabs, bandages, and wound dressings are amoung those considered for enzyme immobilization.

The development of decontamination towelettes as an environmentally benign route to removing nerve agents that are composed of organo-phosphorous compounds has been undertaken through the use of organo-phosphorous hydrolase (OPH) (*20*). The broad substrate specificity and high catalytic turnover rate for organophosphosphorous neurotoxins make OPH-immobilized fabrics a potential remediation for exposoure to compounds like paraoxin (P-O bond hydrolyzed), and Soman and Sarin (P-F bond hydrolyzed). Cotton is a highly absorptive material, and in combination with nerve agent–degrading enzymes, would be appropriate for use as a decontamination towelette. Cotton cellulose has been reported in the use covalent immobilization of organophosphorous hydrolase *(21*). Organophosphorous hydrolase (OPH, EC 3.1.8.1) enzymes were covalently linked to glycine derivatized cotton in a similar manner to that of lyszozyme-immobilized cotton mentioned above. Both carbonyldiimidazole and glutaraldehyde were explored as alternative organic phase and aqueous phase coupling strategies for the linking of OPH to cotton. A comparison of the form activities of OPH preparations coupled in aqueous and organic coupling techniques is shown in Table I. The aqueous coupling with glutaraldehyde demonstrated both better coupling efficiency and better hydrolysis efficiency (as much as 120 hydrolyzed/cm^2 fabric) than the organic

Table I. Comparison of kinetic results of immobilized OPH enzyme on derivatized cotton for Paraoxon hydrolysis[a]

Enzyme Preparation	*Fabric*	*Coupling Method*	*μg Hydrolyzed/min/cm²*
Pure OPH	Mono-gly	Aqueous	39.00
Pure OPH	Mono-gly	Organic	0.47
Pure OPH	Bis-gly	Aqueous	11.00
Crude OPH	Mono-gly	Aqueous	120.00

[a] A comparison of different OPH enzyme preparations within different forms of glycine cotton. The mono-gly is cotton where a single glycine was esterified to cellulose, and the bis-gly is two glycines linked to cotton. The aqueous and organic coupling methods were glutaraldehyde and 1,1,-carbonyldiimidazole as coupling reagents used in buffer and dimethlformamide respectively. The rates of hydrolysis are paraoxon hydrolyzed per minute per square centimeter of cotton fabric.

phase coupling (0.47 hydrolyzed/cm^2 fabric). These activity levels are sufficient to degrade gram-size quantities of nerve agent in a few minutes.

Mechanism Based Wound Dressings

Mechanism based wound dressings are a growing area of interest. This is especially true of chronic wound dressings, and the exponential growth of the dressing market in recent years is, in part, a result of dressing design based on advances in wound healing science. Enzymatic debridement employs both selective and non-selective enzymes that can be topically applied to digest devitalized tissue and jump-start the healing process in the wound bed. Since some enzymes do not distinguish between devitalized and healthy tissue, there is a drawback to this method of debridement that can result in pain to the patient. It has been suggested that a dressing, bearing immobilized proteases that could be removed from the wound bed, would prevent prolonged exposure of healthy tissue to the enzyme. However, a relatively unexplored area of enzymatic debridement is the immobilization of enzymatic debridement proteases on dressings. One exception includes the development of chronic wound debridement wound dressings as wound healing agents using enzymatic debridement utilized in Russia (*22-23*). These dressings employ trypsin. Dalseks-trypsin is a proteolytic ferment form of trypsin which is immobilised on gauze. Dalseks-trypsin has been reported to accelerate healing in pyro-necrotic wounds, burns, and frost-bite in Russia with no side effects. In the western world, two different wound debridement agents are used, that employ the protease enzymes collagenase in a petroleum base and papain and urea in a cream base. Papain is a proteolytic enzyme derived from the fruit of Carica papaya. It is a nonspecific cysteine protease, capable of breaking down a variety of necrotic tissue substrates. Collagenase, derived from *Clostridium histolyticum* is a protease specific for proteolysis which is rich in alkyl side chain amino acids such as proline, leucine, and glycine. It has been suggested that fibrin/fibronectin and collagen are segregated in wounds: fibrin/fibronectin in the upper portion of eschar and collagen in the lower portion of the wound's scab. Thus bacterial collagenase (which is more effective in digesting collagen and elastin) may work from the bottom-up and papain-urea preparations (which are more effective in digesting fibrin) work from the top-down. Based on this, papain might be expected to be more effective as an immobilized enzyme on cotton dressings, but there have been no reports comparing collagnease, papain, trypsin, or other enzymes immobilized on dressings.

Hygenic and Cosmetic Swabs

Applications of immobilized enzymes have also been considered for the cosmetic removal of lipases and proteases from human sebum, in sebaceous glands composed of fatty material and epithelial debris. Fats and cellular protein debris from sebum tend to adsorb dirt and solids causing the skin to flake. The use of immobilized lipases and proteases to break down fats and proteins that cause and contribute to this problem is a potential alternative to soap and detergents which can cause excessive drying and flaking of the skin (*24*). A cosmetic technology of interest in this regard is nonwoven cotton swabs having free and polymer-crosslinked immobilized lipases and proteases with the enzyme-polymer conjugate adsorbed or covalently linked to the cotton fibers. Alternatively, a gel, viscous liquid or encapsulaton within a semipermeable membrane or liposome serves as the delivery vehicle for the enzymes release on the skin.

Other Innovative Uses of Enzymes and Substrates in Relation to Cotton

The study and development of enzyme and enzyme substrate conjugates for specialty and medical applications has been explored from a standpoint of enzyme-polymer conjugates and enzyme substrate-cellulose conjugates. As an example of the types of approaches in relation to cotton, enzyme-polymer resin, enzyme-substrate polymer resin, and enzyme-substrate cellulose conjugates will also be considered here.

Enzyme-Polymer Conjugate: Cellulase-CLEAR for Removal of White Specks on Cotton

Cellulase refers to a family of enzymes which act in concert to hydrolyze cellulose. *Trichoderma reesei* has a cellulase enzyme complex that has been extensively studied. This complex converts crystalline, amorphous, and chemically derived celluloses quantitatively to glucose. A variety of uses of cellulase have been considered, but with cotton fabrics cellulase preparations are used to alter the appearance of their surfaces. Cellulase has been employed to confer a washed appearance in denims, a smooth appearance to fabrics by removing surface fuzz fibers (pills), or to remove tangled bundles of surface fibers that do not pick up dyes and thus result in white specks. SEMs, as shown in Figure 4 of the cotton fiber demonstrate that this phenonmenon termed neps is a flattening of the cotton fiber, which has been associated with development of an immature secondary cell wall that occurs during growth in the field.

In an effort to assess the feasibility of using an immobilized form of cellulase as an effective recyclable treatment for reduction of the white specks found on some dyed fabrics, a cellulase complex was immobilized onto crosslinked ethoxylate acrylate resin (CLEAR) polymer. The synthesis of the

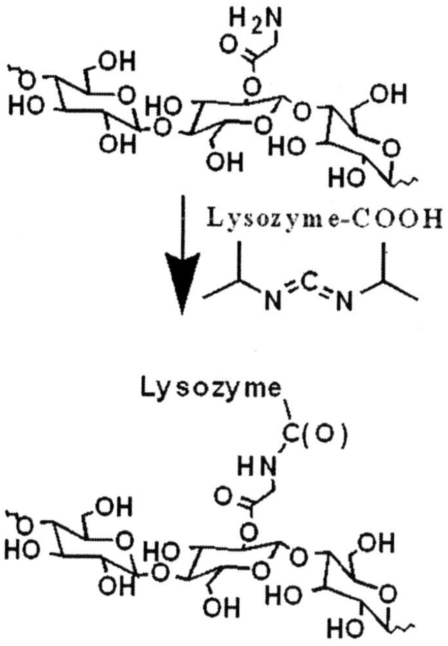

Figure 3. Carbodiimide-mediated reaction of lysozyme with glycine-derivatized cotton cellulose. The reaction proceeds by formation of the O-acyl-isourea adduct of carboxylate amino acid side chains on the protein which then react at the amino group of the derivatized cotton forming a covalent amide bond to link the protein on the surface of the cotton fiber.

cellulase-polymer conjugate was performed similarly to the lysozyme-cellulose conjugate prepared on cotton. This conjugate showed some activity in reducing the white speck count on dyed fabrics as shown in Figure 5.

This research thus demonstrates the feasibility of using immobilized cellulase to reduce white speck counts in dyed fabrics. The manufacturing development and adaptation of the CLEAR-immobilized cellulase is still required for industrial application.

Enzyme Substrate Conjugates of CLEAR and Cellulose: Peptide-Cellulose Conjugate for the Detection of Human Neutrophil Elastase

Enzyme Substrate Conjugates of CLEAR

Although enzyme conjugates have been well studied in the past, little work has been reported on the immobilization and study of enzyme-substrate

FABRIC SURFACE SHOWING A WHITE SPECK NEP,
4 MAGS

Figure 4. Electron microscopic portrayal of the ribbon-like and flattened structure of cotton fibers that is a result of an in complete and immature secondary cell wall of cotton. These structures give rise to the neps or white specks found sometimes on dyed cotton fabrics.

conjugates (a bioconjugate wherein the enzyme's substrate is immobilized), their practical efficacy for use as bio-sensors, and for enzyme detection. In addition to examinig the synthesis and activity of cellulase-CLEAR conjugates as shown above, CLEAR was also applied to examine the behavior of substrate-immobilzed polymers of a peptide substrate of human neutrophil elastase as a detection device for assessment of protease activity in chronic wounds. High levels of human neutrophil elastase (HNE) in chronic wounds have been associated with degradation of cytokine growth factors necessary for normal wound healing. Thus, accurate clinical detection and quantification of HNE is predicted to be important in the future to the therapeutic management of chronic wounds. Colorimetric detection of HNE using a rapid spot test to detect HNE levels on an immobilized enzyme-substrate-derivatized surface could be potentially used at the bed side or imbedded in the dressing design to measure protease levels and thus help control the pathology of protease ridden chronic wounds. The wound dressing designs for lowering HNE levels in the chronic wound have previously been shown (*25*). For this reason an approach for monitoring wound HNE levels in patients with chroinic wounds would enhance the efficiency of assessing the need to apply therapeutic modalities to lower HNE through sequestration or inhibition. The chromogenic peptide substrate Succinyl-Ala-Ala-Pro-Val-pNA (conjugate I) and its analog

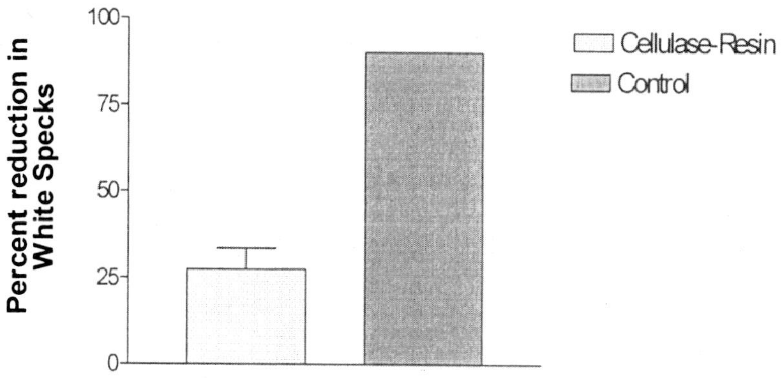

Figure 5. The cellulase-CLEAR conjugate was tested for efficacy by treating white-specked dyed fabrics with the enzyme-polymer under aqueous conditions. White speck were mechanically counted and the percet reduction in white specks was determined relative to CLEAR treatments of dyed fabric without the cellulase.

Succinyl-Ala-Ala-Pro-Ala-pNA (conjugate II) were attached to derivatized CLEAR (*26*). The ability of CLEAR to swell in both organic and aqueous solvents provides the potential for assessment of enzyme active substrates immobilized on resin. The protease turnover activity and enzyme binding of these conjugates was measured.

Three approaches have been used in assessing elastase activity on the CLEAR-resin-substrate conjugates. The initial approach involved assessment of enzyme activity directly on substrate-resin. The relative activities between the immobilized substrate-resin and the substrate in solution showed rates from seven to twenty times faster in solution than with substrate-resin. However, the kinetic profiles of the substrate-resins were minimally measurable under these conditions, due to light scattering of the resin beads. Subsequently the resin-substrate conjugate was saponified and the peptide-polymer ester analogs released into solution. Amino acid analysis revealed nearly quantitative release of the peptide ester analogs. The results of the kinetics and chromophore release for the analogs showed that 0.6-28% of the peptide ester analogs reacted with enzyme and underwent substrate hydrolysis. The peptide ester analog of conjugate I demonstrated that 22-28% of available substrate was hydrolyzed; whereas conjugate II demonstrated an average of 0.6% of available substrate was hydrolyzed. A third approach to estimating the substrate-resin activity was through assessment of the substrate-resin kinetics on a polypropylene surface. The substrate resin was reacted with enzyme on a microtiter-plate polypropylene surface. The amount of chromophore deposited on the polypropylene was found to be 5% of the available substrate. A polymer matrix containing an immobilized elastase substrate with active properties would be adaptable to a 'dip-stick' type of approach for clinical detection and quantification of the enzyme.

The use of immobilized enzyme substrates for the assessment of enzyme activity has received scarce attention compared with the immobilization of enzymes. For example, collagen has been attached to synthetic surfaces as a model to evaluate enzyme-substrate reactions and diffusion with a surface bound substrate. An approach similar to the one discussed in this paper involved chymotrypsin-catalyzed hydrolysis of immobilized substrates using agarose and polyacrylamide-bound substrates of L-phenylalanine 4-nitroanilide (*27*).

Immboilized Substrate-Cellulose Conjugates

A follow-up study with immobilized cellulose conjugates of the chromogenic peptide examined above was undertaken as a practical method for potential clinical assessment of elastase levels in chronic wounds. The structure of the cellulose esterified chromophore is shown in Figure 6. The use of cellulosic paper as an enzyme substrate matrix for visual detection presents an alternative and an efficient method suitable for clinical application in a 'dipstick' format. The colorimetric response of the cellulose-bound chromophore was assessed in HNE buffered solutions by monitoring release of para-nitroaniline from derivatized paper. Both visual and spectral detection methods were used to determine elastase levels from 5.0×10^{-3} to 6.0 units. A comparison of the cellulosic analogs in Figure 7 demonstrated that the cellulose ester of suc-Ala-Ala-Pro-Val-pNA gave stronger absorption than the cellulose ester of suc-Ala-Ala-Pro-Ala-pNA. Clearly this technology is promising, but improved sensitivity is needed to detect lower levels of HNE in chronic wound fluid. Future work will focus on improved chromophores and chemistries of cellulose attachment.

Conclusion

The design, preparation, and assessment of textile fibers containing enzyme conjugates has a promising future. This in part is because the design of new fibers for use in health care textiles has increased dramatically over the last twenty-five years due to advances in polymer, wound, hygiene and textile sciences. The world is also turning to more selective chemistries that function to protect the environment. Enzymes have always been available for this purpose, and cotton fabrics have been subject to new applications since ancient Egypt. The examples listed above are some examples of immobilized enzymes and substrates that are currently in development or on the market. More industrial applications for enzyme conjugates and substrates like these will become available, as enzyme conjugates on textiles are developed for the unmet needs of antimicrobial, decontamination, wound care, hygiene, fabric, and other growing medical and specialty textile markets.

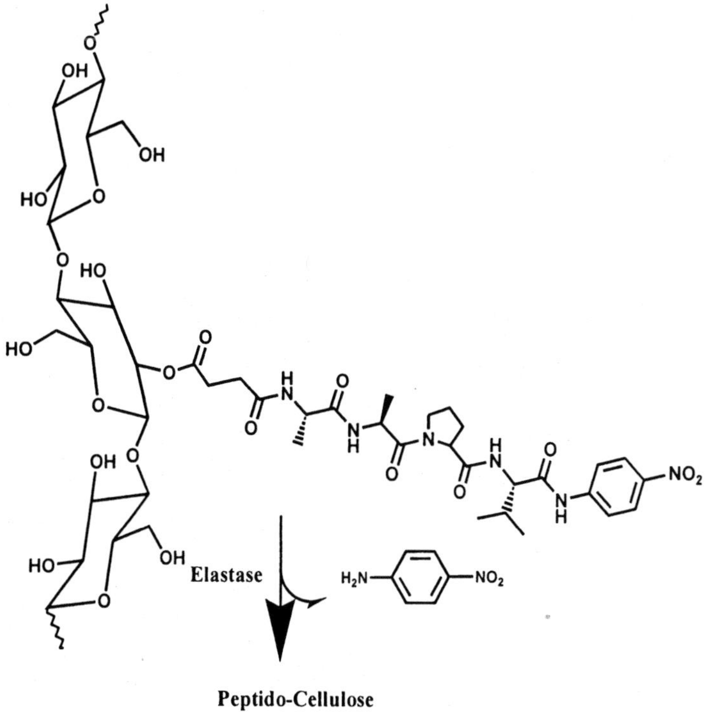

Elastase

Peptido-Cellulose

Figure 6. Portrayal of the chromogenic reaction of the cellulose-substrate conjugate with Elastase. The conjugate containing Succinyl-Ala-Ala-Pro-Val-pNA results in release of the chromophore paranitroaniline which generates a visual response on the paper surface. Elastase is an enzyme that occurs in high concentration in chronic wounds, and may be detected using this type of enzyme hydrolysis with a cellulose-substrate conjugate.

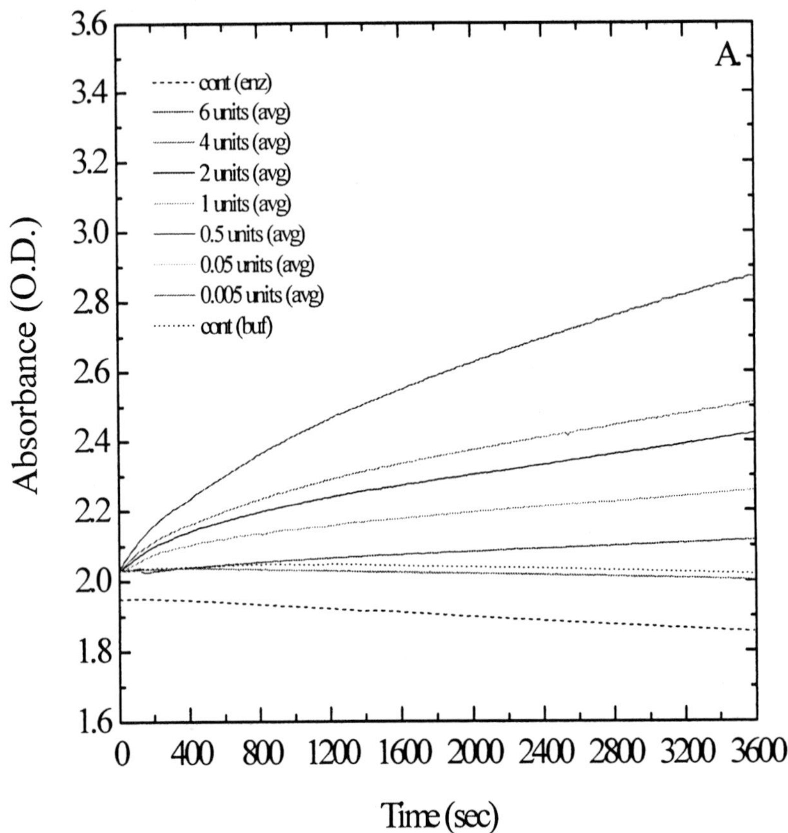

Figure 7. Reaction progress curves of the paranitroanilide peptide conjugate Cellulose-APS-suc-Ala-Ala-Pro-Val-pNA to paranitroaniline was monitored at 405 nm (see figure 6 for reaction) for 1 hr. Hydrolysis of substrate was monitored by combining ~ 5mg of sample with the indicated concentration of elastase. The elastase (Athens) was prepared in a phosphate buffer with a pH of 7.6 (0.2 M sodium phosphate, 0.5 M NaCl, and 6.6% DMSO). HNE units of activity employed in the assay ranged from 5.0×10^{-3} units to 6.0 units. In each well of a 96-well microtiter plate 50 μL of enzyme was reacted with ~5 mg of HNE substrate bound paper. The enzymatic hydrolysis of the immobilized substrate on paper was monitored at 405 nm with incubation at 37 °C using a Bio-Rad Microplate Reader for 1 h. The color that developed on the HNE substrate-bound paper was analyzed by Datacolor International Spectraflash 500. Data was obtained in reflectance mode and converted to absorbance. (See page 9 of color inserts.)

References

1. Albayrak, N.; Yang, S.T. *Biotechnol. Prog.* **2002**, *18(2)*, 240-251.
2. Melo, J. S.; Kubal, B. S.; D'Souza, S. F. *Food. Biotechnol.* **1992**, *6*, 175-186.
3. Krastanov, A. *Appl. Microbiol. Biotechnol.* **1997**, *47*, 476-481.
4. Kamath, N.;D'Souza, S. F. *Enzyme Microb. Technol.* **1992**, *13*, 935-938.
5. Kolot, F. B. *Process Biochem.* **1981**, *5*, 2-9. (32
6. D'Urso, E. M.; Fortier, G. *Enzyme Microb. Technol.* **1996**, *18*, 482-488.
7. Melo, J. S.; D'Souza, S. F. *Appl. Biochem. Biotechnol.* **1992**, *32*, 159-170.
8. D'Souza, S. F.; Melo, J. S.; Deshpande, A.; Nadkarni, G. B. *Biotechnol. Lett.* **1986**, *8*, 643-648.
9. Nakamura, S.; Kato, A.; Kobayashi, K. *J. Agric. Food Chem.* **1991**, *39*, 647-650.
10. Hayashi, K.; Kasumi, T.; Kubo, N.; Haraguchi, K.; Tsumura, N. *Agric. Biol. Chem.* **1989**, *53*, 3173-3177.
11. Scott, D.; Hammer, F. E.; Szalkucki, T. J. In *Food Biotechnology*, D. Knorr; Ed.; Marcel Dekker: New York, **1987**, pp 413-419
12. Moser, I.; Dworsky, P.; Pittner, F. *Appl. Biochem. Biotechnol.* **1988**, *19*, 243-249.
13. Chen, J.-P.; Chen, Y.-C. *Biores. Technol.* **1997**, *60*, 231-237.
14. Chen, J.-P.; Chen, Y.-C. *Biores. Technol.* **1997**, *60*, 231-237.
15. Salton, M. R. J.; Ghuysen, J. M. *Biochim. Biophys. Acta* **1959**, *36*, 552-554.
16. Salton, M. R. J.; Ghuysen, J. M. *Biochim. Biophys. Acta* **1960**, *45*, 355-363.
17. Jeanloz, R. W.; Sharon, N.; Flowers, H. M. *Biochem. Biophys. Res. Commun.* **1963**, *13*, 20-25.
18. Blake, C. C. F.; Mair, G. A.; North, A. C. T.; Phillips; Sarma, V. R. *Proc. R. Soc. Ser. B* **1967**, *167*, 378-388.
19. Edwards, J. V.; Sethumadhavan, K.; Ullah, A. H. J. *Bioconjugate Chem.*, **2000**, *11(4)*, 469-473.
20. Russell, A. J.; Berberich, J. A; Drevon, G. F.; Koepsel, R. R. *Ann. Rev. Biomed. Eng.* **2003**, *5*, 1-27.
21. Grimsley, J.K.; Singh, W.P.; Wild, J.R.; Giletto, A. In *Bioactive Fibers and Polymers*; Edwards, J. V.; Vigo, T. L.; Eds.; ACS Symposium Series 792: Washington D.C., p 35-49.
22. Hogan, M. K. *Healthy Technology Assessment*, **1999**, *3*, 17.
23. Tolstykh, P.I.; Gostishchev, V.K.; Khanin, A.G.; Iusuf, M.I.; Aboiants, R.K. *Technical Textiles Inter*national, **1995**, *4 (4)*, 14-15.
24. Calvo, L.C., *Immobilized Enzymes*, United States Patent, 4556554, 1985.
25. Edwards, J.V. In *Modified Fibers with Medical and Specialty Applications*, Edwards, J. V.; Buschle-Diller, G.; Goheen, S. C.; Eds.; Springer: The Netherlands, **2005**, pp 11-33.
26. Edwards, J.V.; Pierre, S.C.; Bopp, A.F.; Goynes, W. J. *Peptide Research*, **2005**, *66*, 160.
27. Fischer, J.; Lange, L.; Jakubke, H. D. *Eur. J. Biochem.* **1978**, *88*, 453-457.

Chapter 13

New Developments of Enzymatic Treatments on Cellulosic Fibers

Su-yeon Kim, Andrea Zille, Andreia Vasconcelos, and Artur Cavaco-Paulo*

Department of Textile Engineering, University of Minho, Campus de Azurém 4800–058 Guimarães, Portugal

Recent, non-conventional enzyme bioconversions of cellulosic fibers for the textile industry are reviewed. Cellulosic fibers are the largest market of the textile industry, and also represent the most successful market for enzyme based processes in the textile area. The new enzyme developments presented in this book chapter include the use of enzymes to recover strength in resin-crosslinked cellulose fabrics, to phosphorylate cellulose fabrics for improved antiflame retardancy, and to coat and functionalize cellulosic fabrics. Enymes utilized include cellulases, lipases, proteases, and laccases.

Introduction

The use of enzymes in textile processing has been reported since the middle of the 19th century, when malt extract was used to remove amylaceous sizes before dyeing and printing (*1*). Enzymatic processes have been developed for wet processing of textile goods in wide ranging operations from cleaning preparations to finishing processes.

The processing of cellulosic fibers is generally performed under alkaline conditions. This is mainly because the free hydroxyl groups of 1→4-β-D glycosodic units only become ionized at high pH values. In general, all cellulose textile treatments undertaken at high pH's require subsequent neutralization washing steps that involve large amounts of water. The reactive dyeing of cellulosic fibers is a well known example of this type of treatment. The application of biotechnology to process cellulosic fibers could provide a more eco-friendly alternative option for its modifications at milder conditions.

The possibility of replacing the alkaline scouring process of cellulose fabrics with an enzyme bioprocess has been widely studied (*2*). This has included the enzymatic decomposition of non-cellulosic impurities, and increased wettability of the textile material using various types of enzyme-cellulases, pectinases, lipases, and proteases under different applications (*2*). Enzymes have their own conditions of temperature and pH for optimum activity (*1*). Consequently, temperature and pH control are important in enzymatic treatments if the maximum activity of the enzymes is to be obtained. In this book chapter, new enzymatic methods to hydrolyze crosslinks, phosphorylate, and coat and functionalize cellulose fabrics, with the use of several enzymes such as lipases, proteases, cellulases, and laccases, are reviewed and described.

Lipases and Proteases to Improve the Mechanical Characteristics of Durable Press Finished Cotton Fabrics

Durable press finishing processes are widely used in the textile industry to impart wrinkle-resistance to cotton fabrics and garments. However formaldehyde is released during the fixation procedure. Therefore, in recent years many efforts have been made to develop formaldehyde-free crosslinking agents. For this purpose, polycarboxylic acids and N-hydroxymethy acryl amide were chosen as the most promising formaldehyde-free agents. It is well known that a major drawback of any crease-resistance treatment on cotton fabrics, is that the improvement of the dimensional stability and wrinkle resistance is always correlated with severe loss of mechanical strength of treated fabrics (*3*). The fabric strength loss caused by crosslinking is a reversible process and can be restored by chemical hydrolysis (*4*). A milder, low concentration enzymatic process could be used instead on the textile, to reduce the strength loss of crosslinked cotton.

The application of lipase (EC 3.1.1.3) type VII from *Candida rugosa* partially restores the strength loss of the fabrics, due to polycarboxylic crosslinking by hydrolysis of the ester linkages. The lipase treatment improves the tensile strength up to 10%, causing just a 4% of alteration of crease-resistance on the fabrics (*4*). A commercial serine protease (EC 3.4.21.62) from a *Bacillus* microorganism (Alcalase 3.0 T, Novo Nordisk) was applied to recover the tensile strength loss of N-hydroxymethylacrylamide crosslinked cotton fabrics. The hydrolysis, of the amide bond in the cross-linked cellulose by a protease, resulted in ~15% recovery of strength loss coupled with up to 8% reduction of the crease-resistance effect (*5*). In the conventional alkaline hydrolysis, there is a diminution of the wrinkle resistance of the fabrics, as well as a diminution of the strength loss. Most lipases are serine hydrolases, containing a serine residue in their active site, and form an intermediate acyl enzyme substitute complex. The protease hydrolysis of the amide bonds is processed through two steps, acylation and deacylation; the formation of acyl intermediate is the rate-limiting step.

Phosphorylation of Cotton Cellulose

The phosphorus contents of polymers play an important role in the flame retardance properties of fabrics. Phosphorus containing compounds, usually act in the solid phase of burning materials. When heated, the phosphorus reacts to give a polymeric form of phosphoric acid (PO_3^{2-}). Chemical phosphorylation is usually a complicated process, requiring several protection and deprotection steps (*6*). However, use of an enzyme bioprocess can eliminate many of these steps and, therefore, make synthesis more efficient.

The enzymatic application of phosphorus on cellulosic fiber to impart a flame-resistance function has been studied by several researchers. Hexokinase enzymes (EC 2.7.1.1) catalyze phosphoryl transfer from adenosine-5'-triphosphate (ATP) to the 6-hydroxyl group of a number of furanose and pyranose compounds. The new biosynthesis process for the phosphorylation of cotton cellulose with the enzyme hexokinase (Type IV from Baker's yeast, Sigma) in the presence of the phosphoryl donor ATP was studied, and the assumed reaction mechanism is illustrated in Figure 1 (*7*). The enzymatically phosphorylated fabrics showed nearly threefold retarded fire propagation (1.1 cm^2/s), *i.e.*, flame resistance, as compared to untreated cotton samples (3 cm^2/s), when exposed to flames.

Production of Light Weight Polyester Fabrics

Due to fashion trends and requirements, there is a market for light-weight polyester fabrics that cannot be produced by the normal power looms in the textile industry. A possible alternative, is the production of fabrics from polyester blended with another fiber, with the other fiber being subsequently eliminated by chemical treatments. The most common of these fabrics are the polyester blends with cellulosic fibers. In this case, the cellulose is normally removed with a 75% (v/v) sulfuric acid solution. The advantages of the acid hydrolysis of cellulose are that it is a fast and cheap process. The disadvantages are that it is a harsh industrial process, and not environmentally friendly. Thus, the enzymatic hydrolysis process is a cleaner alternative that occurs under the less harsh conditions of atmospheric pressure, moderate temperatures, and mild pH conditions. In nature, cellulose degradation is performed by cellulases of

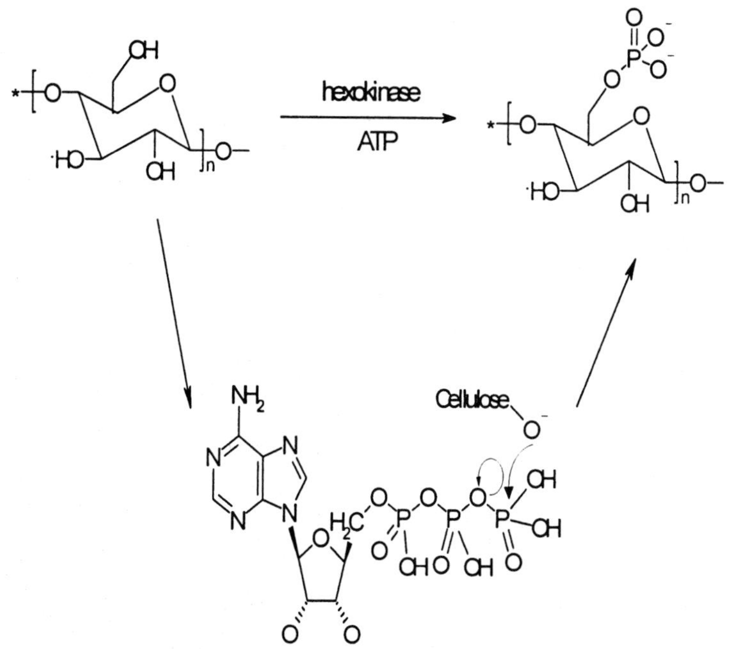

Figure 1. Mechanism of hexokinase-catalysis cellulose phosphorylation (Reproduced with permission from reference 7).

bacterial or fungal origin. The efficient degradation of this substrate requires the combined action of several types of cellulases: endoglucanases (EC 3.2.1.4) that produce new ends randomly within the polysaccharide chain; cellobiohydrolases (EC 3.2.1.91) which release cellobiose units from the cellulose chain ends; and β-glucosidases (EC 3.2.1.21) that convert cellobiose into glucose. Cotton fibers can be completely converted into microfibrilar material by combined action of enzyme hydrolysis and strong mechanical effects in short periods of time. The production of insoluble material is dependent on the level of heating effects present inside the treatment pots (*8*).

The production of light-weight polyester fabrics was investigated by enzymatic removal of the cellulose from a polyester/cotton blended fabric with a commercial cellulose (Cellusoft L, Aquitex). The removal of cotton from the blended fabric yields more than 80% of insoluble microfibrilar material by combined action between high temperature (heating) effects and cellulose hydrolysis. In such a process, it was verified that other major features like bath ratio, enzyme dosage, and treatment time, play an important role on the conversion of cotton fibers into microfibrilar material (*9*).

Laccases Polymer Coatings Over Cellulose

Laccases (EC 1.10.3.2) are multi-copper enzymes that catalyze the oxidation of a wide range of inorganic and organic substances using oxygen as an electron acceptor (*10*). Laccases are able to catalyze the transformation of various aromatic compounds, specifically phenols and anilines, through an oxidative coupling reaction of concomitantly reducing molecular oxygen to water (*11-13*), and the mechanism of enzymatic polymerization is shown in Figure 2. Depending on the 'R-group' different properties can be delivered to the fiber.

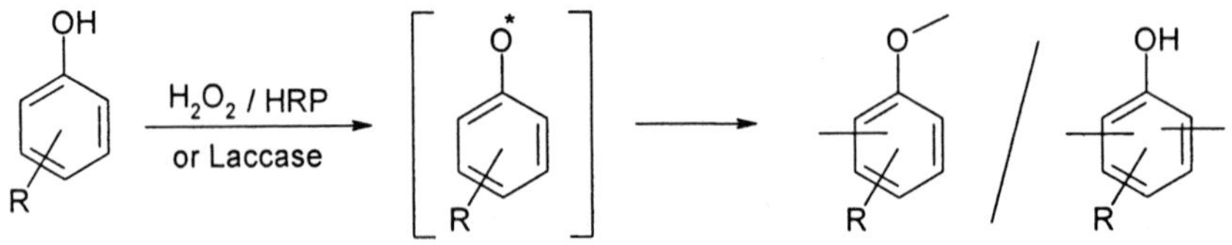

Figure 2. Laccase catalyze polymerization of phenolic compounds (Reproduced with permission from reference 14).

Solution polymerization of phenolic compound by an enzyme in the presence of cellulosic fiber is undertaken to produce polymer coated materials. Many enzymatically polymerized phenolic compounds tend to have a characteristic color because polyphenol forms a large conjugated structure along the main chain (*14*). Laccase from *Trametes hirsuta* can be used in the bleaching process of cellulosic fiber materials. The short time of the enzymatic pre-treatment is sufficient to enhance fabric whiteness, which renders this bio-process suitable for continuous operations. This low-enzyme consuming laccase pre-treatment at mild conditions, can reduce significantly the hydrogen peroxide dosage in subsequent chemical bleaching (*15*).

In the case of cotton coating by laccase, the cotton fabric has been previously dyed and functionalized. Afterward, the formed amino groups on the cotton fabric are coupled with the enzymatic synthetized polymer. The products obtained from the coupling processes of aminized cellulose with products obtained from catechol oxidation by laccase are identified by LC/MS analysis. This mechanism and the products confirmed by LC/MS analysis is shown in Figure 3.

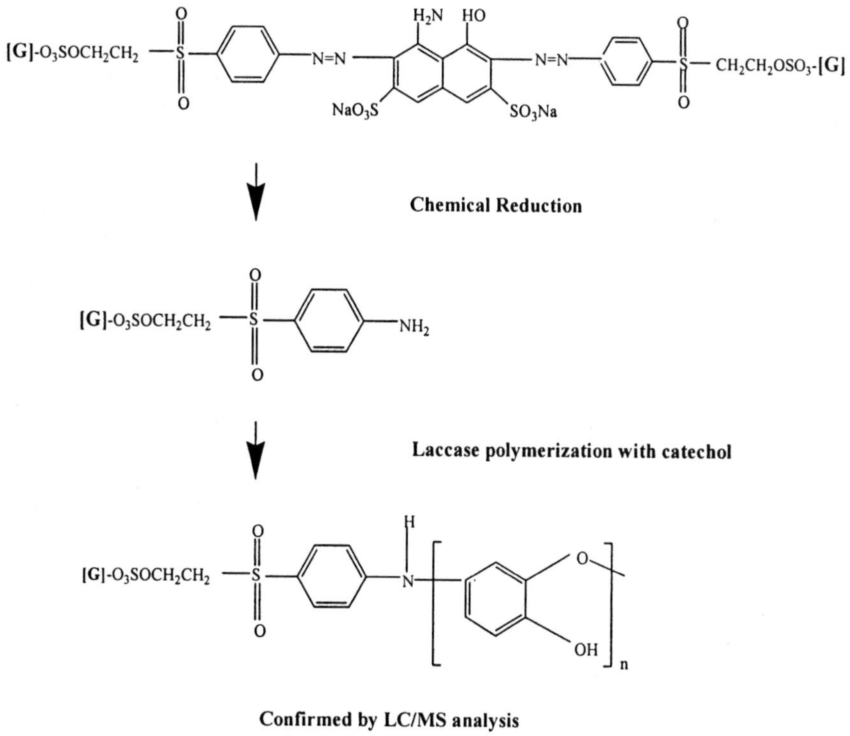

Chemical Reduction

Laccase polymerization with catechol

Confirmed by LC/MS analysis

Figure 3. Chemical mechanism of colorization of cotton cellulose fiber by laccase and the products of reaction were confirmed using LC/MS analysis.

References

1. Aly, A. S.; Moustafa, A. B. *J. Cleaner Production.* **2004**, *12*, 697.
2. Cavaco-paulo, A.; Gubitz, G. In *Textile Processing with Enzymes*; Woodhead Publishing Ltd: Cambridge, England, **2003**; p 72.
3. Sricharussin, W.; Ryo-Aree, W.; Intasen, W.; Poungraksakirt, S. *Text. Res. J.* **2004**, *74*, 475.
4. Tzanov, T.; Stamenova, M.; Betcheva, R.; Cavaco-Paulo, A. *Macromol. Mater. Eng.* **2002**, *287*, 462.
5. Stamenova, M.; Tzanov, T.; Betcheva, R.; Cavaco-Paulo, A. *Macromol. Mater. Eng.* **2003**, *288*, 71.
6. Edwords, J. V.; Yager, D. R.; Cohen, I. K.; Diegelmann, R. F.; Montante, S.; Bertoniere, N.; Bopp, A. F. *Wound Repair Regen.* **2001**, *1*, 50.

7. Tzanov, T.; Stamenova, M.; Cavaco-Paulo, A. *Macromol. Rapid. Commun.* **2002**, *23*, 962.
8. Morgado, J.; Cavaco-Paulo, A.; Rousselle, M. *Text. Res. J.* **2000**, *70*, 696.
9. Vasconcelos, A.; Cavaco-Paulo, A. *Cellulose*. **2006**, in press.
10. Zille, A.; Gornacka, B.; Rehorek, A.; Cavaco-Paulo, A. *Appl. Environ. Microbiol.* **2005**, *71*, 6711.
11. Zille, A.; Munteanu, F.; Gubitz, G.; Cavaco-Paulo, A. *J. Mol. Catal. B: Enzym.* **2005**, *33*, 23.
12. Guresir, M.; Akatas, N.; Tanyolac, A. *Prog. Biochem.* **2005**, *40*, 1175.
13. Gianfreda, L.; Sannino, F.; Rao, M. A.; Bollag, J.-M. *Water Res.*. **2003**, *37*, 3205.
14. Shin, H.; Gubitz, G.; Cavaco-Paulo, A. *Macromol. Mater. Eng.* **2001**, *286*, 691.
15. Tzanov, T.; Basto, C.; Gubitz, G.; Cavaco-Paulo, A. *Macromol. Mater. Eng.* **2003**, *288*, 807.

Basic Research to Underpin Future Advances in the Application of Industrial Enzymes

Chapter 14

Phosphorylases in the Production of Oligosaccharides

Motomitsu Kitaoka

National Food Research Institute, Tsukuba, Ibaraki 305–8642, Japan

Phosphorylases are one of the three main types of enzymes involved in the formation and cleavage of glycosyl linkages. They are expected to function as unique catalysts in the production of oligosaccharides, owing to their reactions being reversible and highly regio-specific. However, they have not been studied as extensively as the other two types, the hydrolases and synthases. Pairs of phosphorylases have been used to prepare oligosaccharides, including trehalose, cellobiose, and laminaribiose. New processes using phosphorylases to produce oligosaccharides with α-1,2 glucosyl linkage and amylose are reported.

Introduction

The enzymes involved in the formation and cleavage of glycosyl linkages can be divided into three main types, hydrolases, phosphorylases, and synthases (Figure 1) (*1*). The hydrolases are by far the most commonly used in industry, being used to hydrolyze polysaccharides, including starch. Because of their transglycosylation activity, hydrolases are also used to produce oligosaccharides commercially. However, the transglycosylation reaction must be performed by a single enzyme, theoretically limiting the products to those with the same linkages as those of the starting materials; and as cheap starting materials, such as starch, sucrose, and lactose are limited, so are the products.

	regio specificity	stability
1. Hydrolases		
Gly-OR + H_2O (55 mol/L!!) —irreversible (cleavage)→ Gly-OH + HOR	?	ok
2. Phosphorylases		
Gly-OR + H_3PO_4 ⇄ (reversible) Gly-OPO_3H_2 + HOR	**ok**	**ok**
3. Synthases		
Gly-OR + NDP ←irreversible (synthesis)— Gly-ONDP + HOR (phospho-diester linlage)	ok	?

Figure 1. The three main types of enzymes involved in the formation and cleavage of glycosyl linkages.

The synthases, specifically glycosyl-nucleotide glycosyltransferases, are biologically important because of their role in the synthesis of carbohydrate chains *in vivo*. The reactions catalyzed by synthases are effectively irreversible because a high-energy glycosyl-nucleotide bond (a phosphate diester linkage) needs to be broken. This makes them relatively costly. In addition, most of the synthases are too unstable to be of practical use in the synthesis of carbohydrate chains *in vitro*.

The phosphorylases have properties somewhere in between those of the hydrolases and synthases. Since water molecules are not involved in their reactions, and as the bond energy of the substrate, a glycosyl-phosphate, is not as high as that of a glycosyl-nucleotide, their reactions are reversible. This makes them relatively cheap and of practical use in the synthesis of oligosaccharides. However, compared with the hydrolases and synthases, they have not been studied extensively. This book chapter describes applications of these enzymes.

Structural Variation of Phosphorylases

Phosphorylases are named by adding "phosphorylase" after the name of the substrate. To date, only 14 phosphorylases have been reported (Table I) (*1-15*) and, structurally, they vary much less than hydrolases and synthases. All 14 phosphorylases catalyze exowise phosphorolysis at the nonreducing-end of the glycosyl linkage to liberate a monosaccharide 1-phosphate. Most of them phosphorolyze glucosyl linkages to form α- or β-glucose 1-phosphate, the others phosphorolyzing a galactosyl linkage, and an *N*-acetylglucosaminyl linkage. The stereo- and regio-specificities of the phosphorylases are very high, phosphorolyzing only specific types of glycosyl linkage. This is useful when the synthesis of oligosaccharides with specific glycosyl linkages is required. Phosphorylases are classified by the anomeric forms of the glycoside phosphorolyzed or the glycosyl 1-phosphate produced. Like the hydrolases, phosphorylases can also be classified according to the anomeric retention or inversion process (Table I). It should be noted that phosphorylases can act on a wide range of substrates, including starch, sucrose, and maltose, which are relatively inexpensive.

Table I. List of phosphorylases

EC 2.4.1.	*Name*	*Anomer*	*Product*	*Ref.*
1	(glycogen) phosphorylase	retaining	α-Glc 1-P	(*2*)
7	sucrose phosphorylase	retaining	α-Glc 1-P	(*3*)
8	maltose phosphorylase	inverting	β-Glc 1-P	(*4*)
20	cellobiose phosphorylase	inverting	α-Glc 1-P	(*5*)
30	β-1,3- oligoglucan phosphorylase	inverting	α-Glc 1-P	(*6*)
31	laminaribiose phosphorylase	inverting	α-Glc 1-P	(*7*)
49	cellodextrin phosphorylase	inverting	α-Glc 1-P	(*8*)
64	trehalose phosphorylase	inverting	β-Glc 1-P	(*9*)
97	β-1,3-glucan phosphorylase	inverting	α-Glc 1-P	(*10*)
211	lacto-*N*-biose phosphorylase	inverting	α-Gal 1-P	(*11*)
216	trehalose 6-phosphate phosphorylase	inverting	β-Glc 1-P	(*12*)
230	kojibiose phosphorylase	inverting	β-Glc 1-P	(*13*)
231	trehalose phosphorylase	retaining	α-Glc 1-P	(*14*)
nd	chitobiose phosphorylase	inverting	α-NAG1-P	(*15*)

Preparation of Various Oligosaccharides

All the phosphorylases are highly specific toward the donor substrate. However, because they are not highly specific toward acceptors, a range of desired oligosaccharides can be synthesized. Oligosaccharides that have been synthesized using phosphorylases are listed in Table II (*16-35*), and some of our research on the topic is described below.

Table II. Preparation of Oligosaccharides by Using a Phosphorylase

Products	*Enzyme Used*	*Ref.*
α-1,4 linked 2-deoxy-glucosides	glycogen phosphorylase	(*16*)
derivatives of maltose	maltose phosphorylase	(*17-20*)
derivatives of sucrose	sucrose phosphorylase	(*21*)
derivatives of cellobiose	cellobiose phosphorylase	(*22-27*)
β-1,4 linked 2-deoxy-glucosides	cellobiose phosphorylase	(*28*)
laminarioligosaccharides	laminaribiose phosphorylase	(*29*)
cellodextrins and their derivatives	cellodextrin phosphorylase	(*30-31*)
β-1,3,1,4 linked oligosaccharides	cellodextrin phosphorylase	(*32*)
β-1,4 linked gluco/xylo oligosaccharides	cellodextrin phosphorylase	(*33*)
α-glycosides of lacto-*N*-biose	lacto-*N*-biose phosphorylase	(*34*)
Glcβ-1,4-GlcNAc	chitobiose phosphorylase	(*35*)

Utilization of Cellobiose Phosphorylase

Cellobiose phosphorylase catalyzes reversible phosphorolysis of cellobiose, but not cellotriose or higher oligosaccharides. With regard to acceptor specificities of cellobiose phosphorylase, we found that the hydroxyls at positions β-1, 3, and 4 in cellobiose phosphorylase from *Cellvibrio gilvus* were essential, while those at positions 2 and 6 were not, allowing various derivatives to be formed (Figure 2). To date, we have prepared various oligosaccharides using cellobiose phosphorylase (Figure 3) (*23-27*).

Recently, we demonstrated that 2^{II}-deoxycellobiose and its derivatives can be formed using cellobiose phosphorylase and glucal as the donor substrate (*28*). Similarly, 2-deoxy-α-glucosides have been formed using glycogen phosphorylase (*16*).

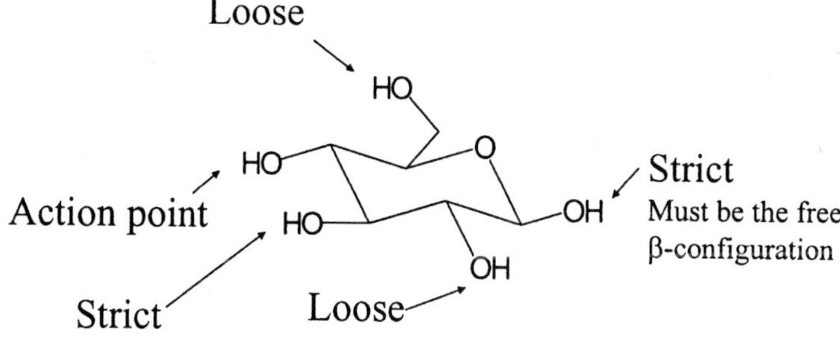

Figure 2. Acceptor specificity of cellobiose phosphorylase from Cellvibrio gilvus.

Synthesis of a Library of β-(1→4) Hetero- Glucose and Xylose-Based Oligosaccharides

Cellodextrin phosphorylase (CDP) can reversibly phosphorolyze the glucosyl residue at the reducing-end of cellooligosaccharides (cellotriose or larger ooligosaccharides) into α-glucose-1-phosphate with a perfect stereo- and regio-specificity. We examined the reverse phosphorolysis of CDP with α-glucose-1-phosphate or α-xylose-1-phosphate as the donor substrates, and cellobiose, xylobiose, Glcβ-1,4-Xyl, or Xylβ-1,4-Glc as the acceptor substrates. We successfully synthesized all 6 hetero-trisaccharides and 10 of the 14 possible hetero-tetrasaccharides (*33*). It was not possible to synthesize the 4 tetrasaccharides with a Xyl-β-1,4-Glc sequence at their non-reducing ends employing this method. The β-1,4 linked glycan chains found in cellulose, xylan, chitin, and chitosan differ only in their constituent monosaccharides (*i.e.*, D-glucose, D-xylose, *N*-acetyl-D-glucosamine, and D-glucosamine, respectively), and are hydrolyzed by the corresponding enzymes, cellulase, xylanase, chitinase, and chitosanase, respectively. The library of the heterosaccharides is considered to be useful for characterizing cellulases and xylanases.

Synthesis of Laminarioligosaccharides

Mixtures of laminarioligosaccharides with varying degrees of polymerization were synthesized from glucose-1-phosphate and glucose by using a cell-free extract of Euglena gracilis containing laminaribiose phosphorylase and β-1,3-oligoglucan phosphorylase (29). The average degree of polymerization increased with an increase in the α-glucose-1-phosphate/glucose ratio. With ratios of 1 and 20, laminarioligosaccharides with degrees of polymerization of 1-9 and 2-14 were detected by HPLC, and the average degrees

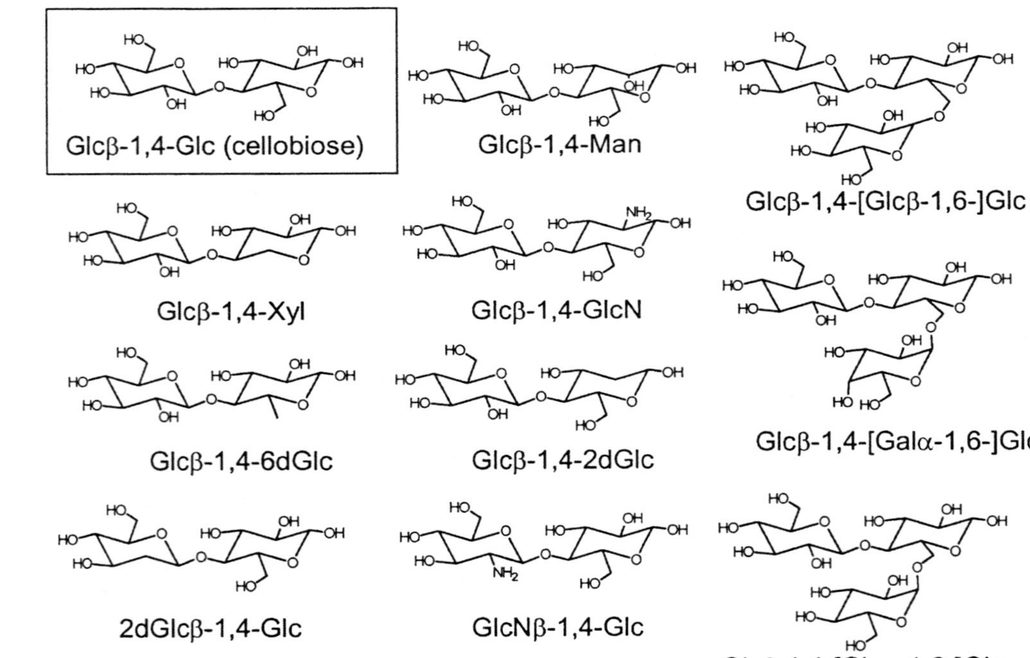

Figure 3. Oligosaccharides synthesized using cellobiose phosphorylase from Cellvibrio gilvus.

of polymerization were 1.8 and 6.3, respectively. Laminarioligosaccharides larger than laminaridecaose, which are difficult to obtain using the standard method involving limited hydrolysis of β-1,3-glucan, were synthesized using this method.

Combination of Two Phosphorylases to Produce Oligosaccharides

Maltose to Trehalose

The first attempt to prepare an oligosaccharide by using a combination of two phosphorylases was reported in 1985 (*36*). Maltose was converted into trehalose by maltose phosphorylase and trehalose phosphorylase (inverting) with a catalytic amount of phosphate (Figure 4). Both phosphorylases phosphorolyzed their substrates into β-glucose-1-phosphate. In 1985, the only possible source of trehalose phosphorylase was *Euglena gracilis*, a protozoan possessing chloroplast, which made industrial application difficult. Later, however, a bacterial trehalose phosphorylase was found (*37-38*) and became important in the industrial of production of trehalose. Trehalose is valuable for its ability to protect against damage by freezing and drying. However, a better process of producing trehalose from starch was later introduced by Hayashibara Co., Japan (*39-40*).

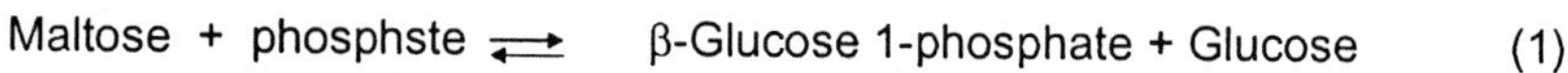

Maltose + phosphste ⇄ β-Glucose 1-phosphate + Glucose (1)

β-Glucose 1-phosphate + Glucose ⇄ Trehalose + Phosphate (2)

Maltose ⇄ Trehalose

(1) Maltose phosphorylase (*Lactobacillus brevis*)
(2) Trehalose phosphorylase (inverting) (*Euglena gracilis*)

Figure 4. Reaction scheme to produce trehalose from maltose using two phosphorylases.

Sucrose to Cellobiose

We demonstrated the direct conversion of sucrose into cellobiose using three enzymes (*1,41*). Cellobiose can be synthesized from glucose-1-phosphate and glucose can be synthesized using cellobiose phosphorylase, but this route is not good because glucose 1-phosphate is prohibitively expensive. Sucrose phosphorylase can be used to generate glucose 1-phosphate. Thus, a reaction system to form cellobiose and fructose using sucrose and glucose as the raw materials along with a catalytic amount of phosphate can be constructed using the combined action of sucrose phosphorylase and cellobiose phosphorylase. However, this reaction system still has two problems: (i) the formation of an equimolar amount of fructose as a byproduct and, (ii) the competitive inhibition by glucose toward glucose-1-phosphate for both enzymes (competitive substrate inhibition for cellobiose phosphorylase [*42*]) that causes a considerable decrease in the activities of the two enzymes. To solve these problems, a third enzyme, xylose isomerase, was added to the system. Xylose isomerase is used industrially to produce high fructose corn syrup by isomerization of glucose into fructose. The following reactions were anticipated in a reaction mixture containing sucrose, a catalytic amount of phosphate, the three enzymes, sucrose phosphorylase, xylose isomerase and cellobiose phosphorylase: initial phosphorolysis of sucrose into glucose-1-phosphate and fructose by sucrose phosphorylase; subsequent conversion of fructose into glucose by xylose isomerase; and generation of cellobiose and phosphate from glucose-1-phosphate and generation of glucose by cellobiose phosphorylase. Since the phosphate is recycled in the sucrose phosphorylase reaction, sucrose was expected to be converted into cellobiose in a single step (Figure 5). It was found, however, that cellobiose could be synthesized from sucrose in the presence of a catalytic amount of phosphate, as expected. The equilibrium is very favorable for the synthesis of cellobiose, probably because of the high energy of the β-fructofuranosyl linkage in the sucrose molecule. We also found that the generated cellobiose auto-crystallized in the reaction mixture when the starting concentration of sucrose was 500 g/L or more. Since it is possible to separate the cellobiose produced from the reaction mixture by exploiting this phenomenon, a semi-continuous production of cellobiose using this system is possible (Figure 6).

The three-enzyme system can produce another glucobiose if the cellobiose phosphorylase is replaced by another enzyme; trehalose (*43*) and lamianribiose (*44*) have been produced by using trehalose phosphorylase (retaining) and laminaribiose phosphorylase, respectively.

Conclusion

Phosphorylases can be used to prepare oligosaccharides, but not much research has been undertaken on these unique types of enzymes. Their reaction mechanisms, however, have been studied both kinetically and structurally (*45-*

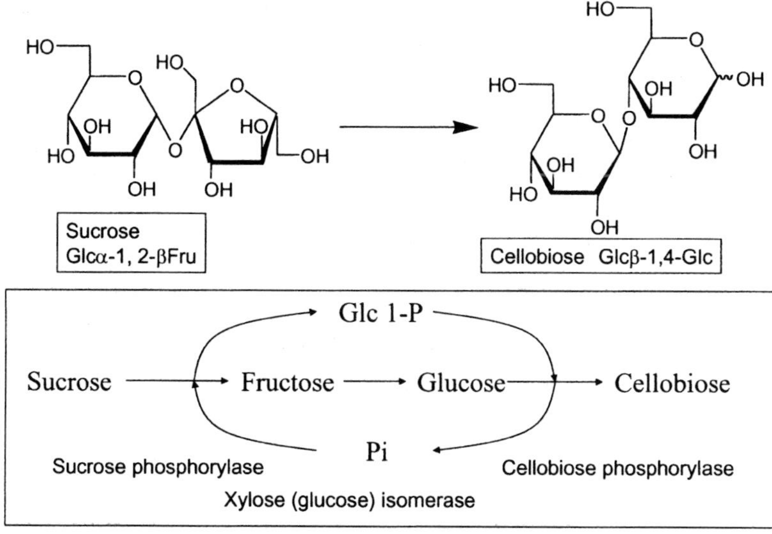

Figure 5. Reaction scheme to produce cellobiose from sucrose using three enzymes.

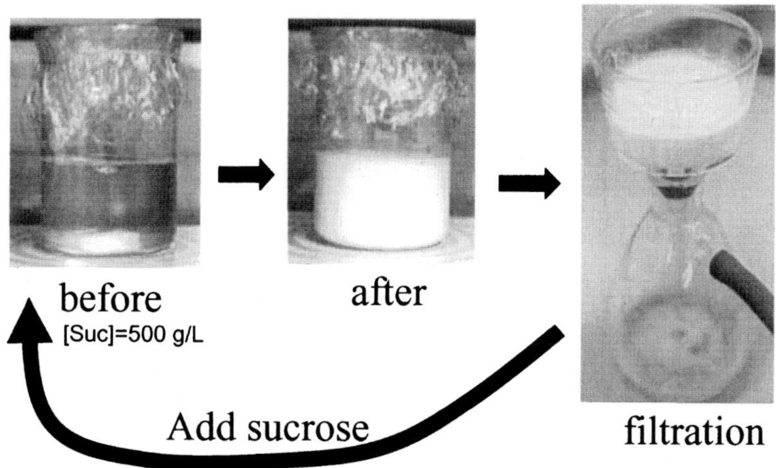

Figure 6. Semi-continuous process to produce cellobiose from sucrose. Cellobiose produced was automatically crystallized in the reaction mixture and separated by filtration.

49). Recently, two possible industrial processes utilizing phosphorylases were reported in Japan. One is the preparation of oligosaccharides with α-1,2 glucosyl linkage, using a combination of inverting trehalose phosphorylase and kojibiose phosphorylase with trehalose as the substrate (*50*). The other is the production of amylose that is fairly homogeneous in the degree of polymerization from sucrose, using sucrose phosphorylase and glycogen phosphorylase (Ezaki Glico Co., Japan). The author hopes that the industrial applications of the phosphorylases will become popular in the near future.

References

1. Kitaoka, M.; Hayashi, K. *Trends Glycosci. Glycotechol.* **2002**, *14*, 35-50.
2. Graves, D. J.; Wang, J. H. In *The Enzymes, 3rd Edn.*; Boyer, P. D.; Ed.; Academic Press: New York, **1972**; *Vol. 7*, pp 435-482.
3. Doudoroff, M. *J. Biol. Chem.* **1943**, *151*, 351-361.
4. Fitting, C.; Doudoroff, M. *J. Biol. Chem.* **1952**, *199*, 153-163.
5. Sih, C. J.; McBee, R. H. *Proc. Montana Acad. Sci.* **1955**, *15*, 21-22.
6. Marechal, L. R. *Biochim. Biophys. Acta* **1967**, *146*, 417-430.
7. Goldemberg, S. H.; Marechal, L. R.; De Souza, B. C. *J. Biol. Chem.* **1996**, *241*, 45-50.
8. Sheth, K.; Alexander, J. K. *J. Biol. Chem.* **1969**, *244*, 457-464.
9. Marechal, L. R.; Belocopitow, E. *J. Biol. Chem.* **1972**, *247*, 3223-3228.
10. Manners, D. J.; Taylor, D. C. *Arch. Biochem. Biophys.* **1967**, *121*, 443-451.
11. Derensy-Dron, D.; Krzewinski, F.; Brassart, C.; Bouquelet, S. *Biotechnol. Appl. Biochem.* **1999**, *29*, 3-10.
12. Andersson, U.; Levander, F.; Rådström, P. *J. Biol. Chem.* **2001**, *276*, 42707-42713.
13. Chaen, H; Nishimoto, T.; Yamamoto, T.; Nakada, T.; Fukuda, S.; Sugimoto, T.; Kurimoto, M.; Tsujisaka, Y. *J. Appl. Glycosci.* **1999**, *46*, 129-134.
14. Kitamoto, Y.; Akashi, H.; Tanaka, H.; Mori, N. *FEMS Microbiol. Lett.* **1988**, *55*, 147-150.
15. Park, J. K; Keyhani, N. O.; Roseman, S. *J. Biol. Chem.* **2000**, *275*, 33077-33083.
16. Evers, B.; Mischnick, P.; Thiem, J. *Carbohydr. Res.* **1994**, *262*, 335-341.
17. Palleroni, N. J.; Doudoroff, M. *J. Biol. Chem.*, **1956**, *219*, 957-962.
18. Kitao, S.; Sekine, H. *Biosci. Biotechnol. Biochem.* **1992**, *56*, 2011-2014.
19. Kitao, S.; Ariga, T.; Matsudo, T.; Sekine, H. *Biosci. Biotechnol. Biochem.* **1993**, *57*, 2010-2015.
20. Kitao, S.; Sekine, H. *Biosci. Biotechnol. Biochem.* **1994**, *58*, 38-42.
21. Selinger, Z.; Schramm, M. *J. Biol. Chem.* **1961**, *236*, 2183-2185.
22. Alexander, J. K. *Arch. Biochem. Biophys.* **1968**, *123*, 240-246.

23. Kitaoka, M.; Taniguchi, H.; Sasaki, T. *Appl. Microbiol. Biotechnol.* **1990**, *34*, 178-182.
24. Tariq, M. A.; Hayashi, K.; Tokuyasu, K.; Nagata, T. *Carbohydr. Res.* **1995**, *275*, 67-72.
25. Tariq, M. A.; Hayashi, K. *Biochem. Biophys. Res. Commun.* **1995**, *214*, 568-575.
26. Percy, A.; Ono, H.; Watt, D.; Hayashi, K. *Carbohydr. Res.* **1998**, *305*, 543-548.
27. Percy, A.; Ono, H.; Hayashi, K. *Carbohydr. Res.* **1998**, *308*, 423-429.
28. Kitaoka, M.; Nomura, S.; Yoshida, Y.; Hayashi, K. *Carbohydr. Res.* **2006**, *341*, 545-549.
29. Kitaoka, M.; Sasaki, T.; Taniguchi, H. *Agric. Biol. Chem.* **1991**, *55*, 1431-1432.
30. Samain, E; Lancelon-Pin, C.; Ferigo, F.; Moreau, V.; Chanzy, H.; Heyraud, A.; Dringuez, H. *Carbohydr. Res.* **1995**, *271*, 217-226.
31. Kawaguchi, T.; Ikeuchi, Y.; Tsutsumi, N.; Kan, A.; Sumitani, J.-I.; Arai, M. *J. Ferment. Bioeng.* **1998**, *85*, 144-149.
32. Moreau, V.; Viladot, J.-L.; Samain, E.; Planas, A.; Driguez, H. *Bioorg. Medic. Chem.*, **1996**, *4*, 1849-1855.
33. Shintate, K.; Kitaoka, M.; Kim, Y.-K.; Hayashi, K. *Carbohydr. Res.* **2003**, *338*, 1981-1990.
34. Farkas, E.; Thiem, J.; Krzewinski, F.; Bouquelet, S. *Synlett*, **2000**, 728-730.
35. Honda, Y.; Kitaoka, M.; Hayashi, K. *Biochem. J.* **2004**, *377*, 225-232.
36. Murao, S.; Nagano, H.; Ogura, S.; Nishino, T. *Agric. Biol. Chem.* **1985**, *49*, 2113-2118.
37. Kizawa, H.; Miyagawa, K.-I.; Sugiyama, Y. *Biosci. Biotechnol. Biochem.* **1995**, *59*, 1908-1912.
38. Aisaka, K.; Masuda, T. *FEMS Microbiol. Lett.* **1995**, *131*, 47-51.
39. Nakata, T.; Maruta, T.; Tsusaki, T.; Kubota, M.; Chaen, H.; Sugimoto, T.; Kurimoto, M.; Tsujisaka, Y. *Biosci. Biotechnol. Biochem.* **1995**, *59*, 2210-2214.
40. Nakata, T.; Maruta, T.; Mitsuzumi, H.; Kubota, M.; Chaen, H.; Sugimoto, T.; Kurimoto, M.; Tsujisaka, Y. *Biosci. Biotechnol. Biochem.* **1995**, *59*, 2215-2218.
41. Kitaoka, M.; Sasaki, T.; Taniguchi, H. *Denpun Kagaku* **1992**, *39*, 281-283.
42. Kitaoka, M.; Sasaki, T.; Taniguchi, H. *J. Biochem.* **1992**, *112*, 40-44.
43. Saito, K.; Kase, T.; Takahashi, E.; Horinouchi, S. *Appl. Environ. Microbiol.* 1998, *64*, 4340-4345.
44. Kitaoka, M.; Sasaki, T.; Taniguchi, H. *Oyo Toshitsu Kagaku* **1993**, *40*, 311-314.
45. Eis, C.; Nidetzky, B. *Biochem. J.* **1999**, *341*, 385-393.
46. Eis, C.; Watkins, M.; Prohaska, T.; Nidetzky, B. *Biochem. J.* **2001**, *356*, 757-767.

47. Nidetzky, B.; Eis, C. *Biochem. J.* **2001**, *360*, 727-736.
48. Eis, C.; Nidetzky, B. *Biochem. J.* **2002**, *363*, 335-340.
49. Hidaka, M.; Honda, Y.; Kitaoka, M.; Nirasawa, S.; Hayashi, K.; Wakagi, T.; Shoun, H.; Fushinobu, S. *Structure* **2004**, *12*, 937-947.
50. Yamamoto, T.; Mukai, K.; Maruta, K.; Watanabe, H.; Yamashita, H.; Nishimoto, T.; Kubota, M.; Chaen, H.; Fukuda, S. *J. Biosci. Bioeng.* **2005**, *100*, 343-346.

Chapter 15

Paradigm for Improving the Catalytic Ability of Industrial Enzymes: Linkage Distortions of Carbohydrates in Complexes with Crystalline Proteins

Alfred D. French and Glenn P. Johnson

Cotton Structure and Quality Research Unit, Southern Regional Research Center, Agricultural Research Service, U.S. Department of Agriculture, New Orleans, LA 70124

Future innovations in applications of industrial enzymes to carbohydrates will require improved knowledge of the mode of action. One aspect of enzymatic hydrolysis of saccharides could be a twisting distortion of the bonds between adjacent monosaccharide residues in carbohydrate substrates. If such twists are important, then new enzymes could be engineered that would increase the distortion for faster reaction. One way to learn if such distortion occurs is to survey existing crystalline carbohydrate-protein complexes. Unusual twists of linkages at the active site in an enzyme may result from catalysis. A related question is whether twisting of the linkage bonds increases the molecular potential energy. For the present work we have tracked the twisting in hundreds of protein–carbohydrate structures and used improved computer modeling of cellobiose to obtain energies. The largest apparent distortions were in similar molecules based on lactose.

Introduction

Emil Fischer's 1894 proposal (*1*) that enzymes and substrates form a "lock and key" relationship was an important step in the understanding of the mode of action of enzymes on their substrates, including carbohydrates. Further thinking resulted in Koshland's 1958 "induced fit" hypothesis (*2*), in which both enzyme and substrate were likely to change shapes as part of the overall reaction mechanism in question. In most situations, neither the substrates nor the enzymes were as rigid as originally envisioned. Work in 2001 by Bosshard (*3*) suggests that "induced fit" is not always entirely applicable for understanding interactions between enzymes and substrates that do not initially have complementary geometries. Instead, some combination of induced fit and conformational selection would be a better description of the interaction. At present, there is ample evidence of large-scale molecular motions that occur in some enzymes (*4*), and there is even a Database of Macromolecular Movements, http://www.molmovdb.org/ (*5*). However, detailed study of changes in shape of carbohydrate substrates has not been widely undertaken.

In a 1966 landmark paper on the crystal structure of a complex of lysozyme and a pentasaccharide, Phillips wrote, "...the concerted influence of two amino acid residues, together with a contribution from the distortion to sugar residue D... is enough to explain the catalytic activity" (*6*). Phillips also stated that such ideas were not novel; that "...activation... by distortion has long been a favorite idea of enzymologists." Distortions could assist in enzymatic reactions by simply elevating the energy, thus putting the structure closer to the activation barrier. However, changed geometry might also make the substrate more vulnerable to attack, or changes in electronic structure resulting from geometric distortion might somehow facilitate the reaction. One elegant proposal suggested the importance of "orbital steering" that would align the parts of the molecules for reaction despite some energetic costs (*7*). At present, we are trying to understand whether types of distortion other than the monosaccharide ring geometry are identifiable and important. Knowledge of such details of the mechanism is essential to engineering enzyme proteins that could, for example, convert cellulose into liquid fuels more rapidly and less expensively. Deliberate changes in the amino acid make-up of an enzyme could cause more distortion (or less!), in such a way that would possibly expedite the reaction.

Distortion of molecules such as oligosaccharides could occur in several ways. In Phillips' paper (*6*), the distortion of the ring shape of residue D was obvious because it no longer had the chair form that is characteristic for such monosaccharides. Another type of distortion would be the internal twisting of a molecule about its inter-residue bonds. Less is known about what would constitute an abnormal, high-energy twist, and what, instead, is just another low-energy molecular shape. This work considers a wide range of experimentally observed geometries of oligosaccharide substrates in crystalline complexes with proteins to see what is normal and what is unusual. We also emphasize calculated potential energies for the twisting through Ramachandran (*8-9*) mapping.

Distortion of φ and ψ Torsion Angles

The linkage torsion angles φ and ψ for β-cellobiose are illustrated in Figure 1 (three-bond linkages are not included here). Although most stick-type physical molecular models permit unhindered rotation about single bonds that are not in rings, the idea of a variation in energy during such rotations is well known. Most chemical scientists are familiar with the preferred staggered and high-energy eclipsed conformations of simple molecules such as ethane or *n*-butane. For *n*-butane, the eclipsed form with all of the carbon atoms coplanar ($\varphi_{C1\text{-}C2\text{-}C3\text{-}C4}$ = 0°) is about 5 kcal/mol higher in energy (*10*) than the anti form ($\varphi_{C1\text{-}C2\text{-}C3\text{-}C4}$ = 180°). Many workers have attributed the preference for the anti geometry to simple steric effects, but the effects of electron delocalization are now thought to be important in determining the energetics of torsional rotation.

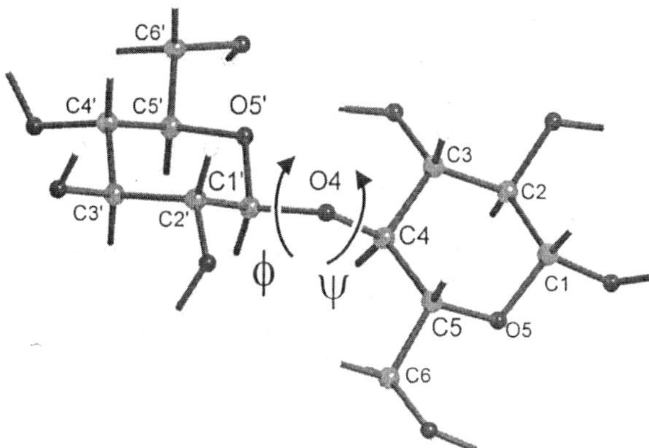

Figure 1. β-Cellobiose from its crystal structure (11) showing the linkage torsion angles φ and ψ, as well as the numbering of carbon and ring and linkage oxygen atoms. We define φ and ψ (here $\varphi_{O5'\text{-}C1'\text{-}O4\text{-}C4}$ = −76.3° and $\psi_{C5\text{-}C4\text{-}O4\text{-}C1'}$ = −132.3°) based on the ring atoms O5' and C5 instead of the commonly used H1' and H4 atoms because protons are not found experimentally in most studies of proteins, and are not well located even in many studies of small molecule crystals.
(See page 10 of color inserts.)

Because carbohydrates are relatively complex molecules, the values of φ and ψ that would correspond to the ground state structures for these isolated molecules are not particularly obvious. Consider the carbohydrate fragment shown in Figure 2 nested in the E1 endoglucanase. Is it clear whether there is a distortion of the tetrasaccharide that would raise its energy in a meaningful way?

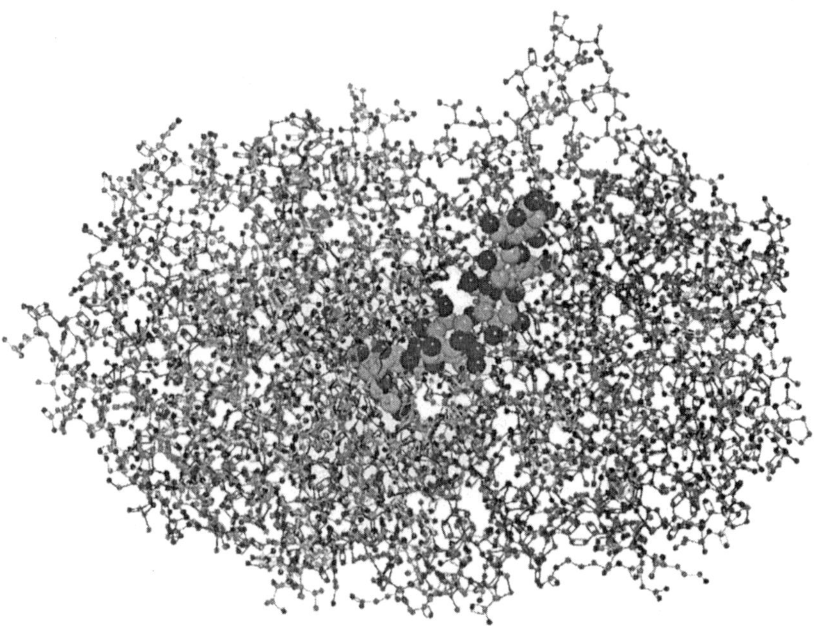

Figure 2. The endocellulase E1 catalytic domain (balls and sticks) with a complexed cellotetraose molecule (space-filling model) from the 1ECE crystal structure (12). The 1ECE designation is an ID code from the Protein Data Bank. Each entry has its own ID code. See (13). (See page 10 of color inserts.)

Both intramolecular hydrogen bonding and stereoelectronic effects play important roles in determining the intrinsic energy in the gas phase, and the lowest energy form results from an often delicate balancing of all the different forces. One approach to determining the ideal conformation is to study all of the appropriate small molecule crystal structures. For example, the normal linkage conformation of fragments of cellulose can be inferred from the linkage conformations of molecules such as cellobiose and lactose and their derivatives (*14*). In that work, the small molecule structures were quite consistent with the known geometries of cellulose and its derivatives. Helical shapes having from 2 to 3 residues per turn resulted from extrapolation of the exact geometries determined by crystallography on small-molecule crystals. Similar results have been reported from fiber diffraction experiments on cellulose and its derivatives and complexes for more than 60 years (*15-16*).

Studies of whether torsion angles are especially distorted by an enzyme could be on surveys of the torsion angles in the crystalline complexes to see if φ and ψ are within the reported range of accurately determined small molecules. Such an analysis is depicted in Figure 3. This is too limited, however, for several reasons. One is that the number of observed structures is finite, and the range could expand with the next crystal structure to be reported. Questions regarding the influence of packing forces will continue, as will the role of water or other solvents during the crystallization process. Carbohydrates also make non-catalytic complexes with proteins, and the effects of that general environment on molecular shape are not widely discussed.

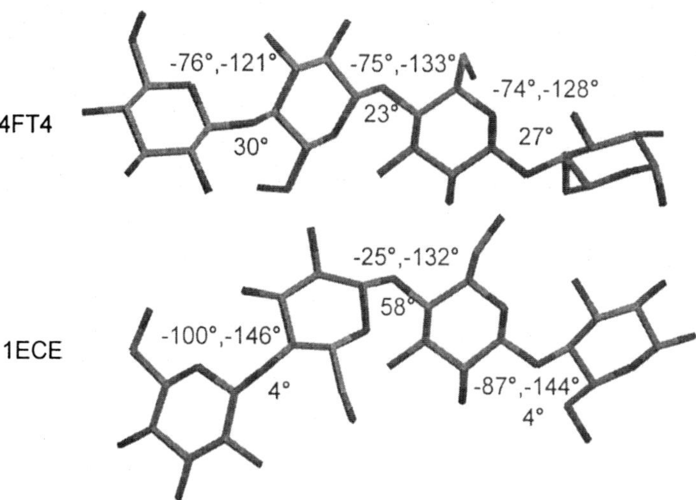

Figure 3. Comparison of cellotetraose fragments from the Protein Data Bank. Paired numbers are the φ and ψ values and the lower, single numbers are the distances in φ,ψ space to the screw axis line (Figures 4-6). The upper model, from the 4TF4 endo/exocellulase crystal structure (17) has φ and ψ values very similar to those of crystalline cellobiose, with an average deviation of 27 degrees from 2-fold screw axis pseudosymmetry. The lower fragment, from the 1ECE endocellulase complex (12), has first and third linkage geometries very close to the frequently observed 2-fold symmetry, but the central linkage is quite different.

(See page 11 of color inserts.)

Because of these limitations, it is worthwhile to augment experimental knowledge of the linkage geometries with theoretical calculations. Other workers have taken a similar approach. Miyake *et al.* (*18*) wrote, "Rotation of a glucosidic bond positions a lone pair of the O4' atom in the direction of Glu172 (which is) acting as a proton donor" (*18*). However, there are many new problems associated with the theoretical work, and at present those results are subject to continuing improvement. Because the computed distortion energy values are the essence of the work, their accuracy is paramount.

Energies

What levels of distortion energy are interesting? One guide might be the energy to distort a sugar ring to give a skew form instead of the usual chair, as observed in Phillips' work. Our previous work (*19*) has given MM energies of about 6 kcal/mol for the chair → skew transformation for glucose. Based on a Boltzmann distribution at room temperature, structures 3 kcal/mol higher than the minimum should occur less than one percent of the time unless there are special, non-random distortions.

Computed energy values for conformations of atomistic molecular models are usually based on one of two different methods. One approach, long named molecular mechanics (MM), is based on empirical force fields. For a given set of atomic positions, the MM potential energy is a summation of contributions from bond stretching, angle bending, van der Waals interactions, electrostatics, and torsion-angle twisting. In the interests of speedy calculations, some force fields go no further. Others add cross terms or special hydrogen bonding functions with the goal of higher accuracy.

Alternatively, energies can be based on electronic structure theory, also called quantum mechanics (QM). In such calculations, energy is based on the spatial distribution of electrons. Naively, electrons would be distributed in the basic orbitals (*e.g.*, *1s*, *2s* and *2p* for atoms in simple carbohydrates), but significant improvements in accuracy result from allowing the electrons to have a wider range of possible locations that are available with basis sets that contain more orbitals. One advantage of QM is that it inherently accounts for the changes in energy that result from redistribution of electrons when the conformation changes, such as in the anomeric effect (*20*). Also, the results depend more on a few fundamental constants such as the charge on electrons rather than extensive parameterization based on experimental (or QM) studies to provide many different constants for empirical equations. That said, QM results for carbohydrates can be quite variable, depending on the basis set and particular method (*21*).

With either MM or QM, most modeling software can relax, or continually adjust, the atom positions to obtain the lowest energy. This is important because it removes the assumption of a particular geometry from the calculations. Without such geometry optimization, models of sucrose that were based on a rigid geometry from one crystal structure had their energies increased by almost 100 kcal/mol when the linkage torsions were twisted to another observed shape (*22*). Our subsequent work with relaxed geometries showed the same twisted structure to be only about 4 kcal/mol above the initial structure (*23*). However, QM also must reallocate and optimize the distribution of electrons for each set of atom positions during the iterative geometry optimization, compounding the time required to complete the calculation. Thus, QM calculations take much longer, a ratio of months to the minutes needed by MM. While computer speeds continue to improve, those gains are often countered with the use of larger basis sets and more complete theory. Improved software and theory are helping to reduce the time.

Ramachandran Maps

An informative way to depict distortion energies is to plot a Ramachandran energy surface or map. The φ and ψ values are systematically varied (typically in increments of 20°) and at each increment the relative energies are noted. Then a contour plot is generated. For 360° ranges of φ and ψ, this means $18\times18 = 324$ energy minimizations. For a carbohydrate such as cellobiose, there are numerous ways that the 8 rotatable hydroxyl groups and 2 primary alcohol groups can be arranged. If each rotatable group could be in any of 3 staggered orientations there would be 3^{10} or 59,043 different combinations. Without a way to know in advance the arrangement that will provide the lowest energy, many "starting structures" must be tried at each φ,ψ increment. Only in that way can the final contoured surface indicate the lowest possible energy at each point. At present levels of performance, a QM map for a disaccharide is based on a serious commitment of resources. Our recent QM map for cellobiose (*24*) took years of computer time despite only covering one fourth of φ,ψ space.

In the present work, we show a map made with our simple hybrid method (*23*) that was important in modeling sucrose. We are updating an earlier study of both cellobiose and maltose linkage geometries (*25*). In this hybrid method, QM is used to make a map for a simple analog of the disaccharide that contains two tetrahydropyran (THP) rings linked with a "glycosidic" oxygen. An MM map for the same analog is made and subtracted from an MM map of the complete disaccharide. Then, the QM analog map is added to the difference. This process provides the critical values for the torsional energies of φ and ψ from QM calculations, but uses MM for the varied rotations of the exo-cyclic groups, saving a great deal of computer time. Compared to our earlier work (*25*), the hybrid map shown in Figures 4-6 is based on a better level of QM and the newer MM4 empirical force field program instead of MM3. The dielectric constant was set to 7.5, minimizing the strength of hydrogen bonding.

Figure 4 shows the map for cellobiose, along with the φ and ψ values that have been found in small molecule crystals such as cellobiose, methyl cellotrioside, cellotetraose, methyl cellobioside, cellobiose acetate, and lactose. Several of the structures are found along the diagonal line that corresponds to structures having 2-fold screw-axis symmetry that is proposed for the most prevalent form of cellulose in higher plants (*26*).

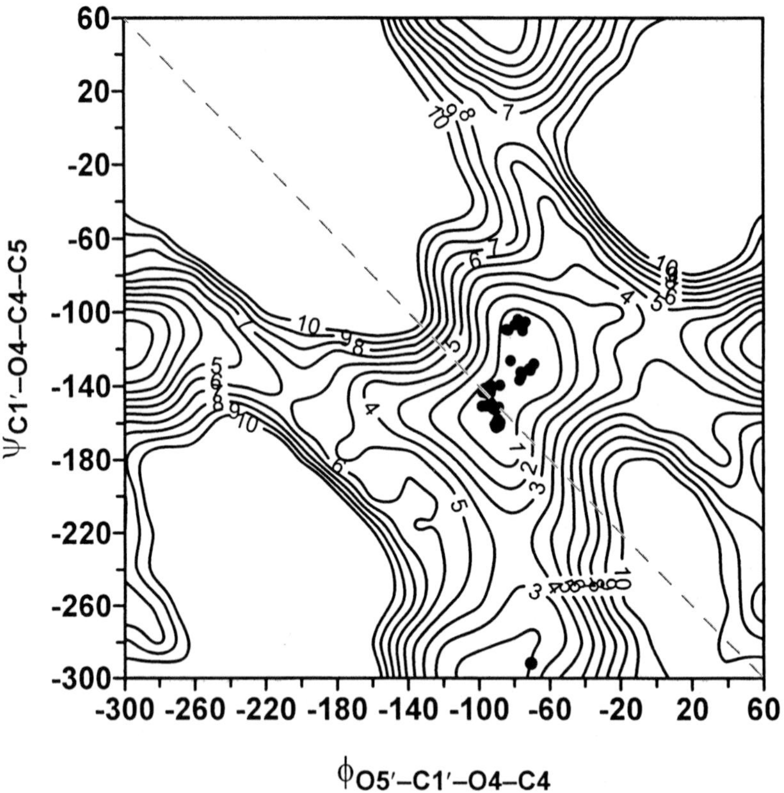

Figure 4. B3LYP/6-311++G(d,p)::MM4 hybrid plot for cellobiose, with energy contours greater than 10 kcal/mol not shown. The dashed diagonal line indicates 2-fold screw-axis symmetry, and the black circles indicate observed conformations in small-molecule crystal structures.

Unlike cellulose, the small molecules cannot have actual internal 2-fold screw symmetry because the reducing and non-reducing ends are different. (In cellulose the ends are so far apart that the lack of symmetry is not resolved by diffraction.) Besides the group of structures near the 2-fold line, a second group, intermediate in distance from the line, contains β-cellobiose. A third group, furthest from the line, is composed of structures that do not have the intra-molecular, inter-residue O3—O5′ hydrogen bond (*27*). Finally, there is a lone conformation at the bottom of the map. Except for that point, all structures are within the 1-kcal/mol contour line. Unlike many of our earlier energy surfaces for cellobiose and its analogs, the line corresponding to the 2-fold axis passes through the 1-kcal/mol contour.

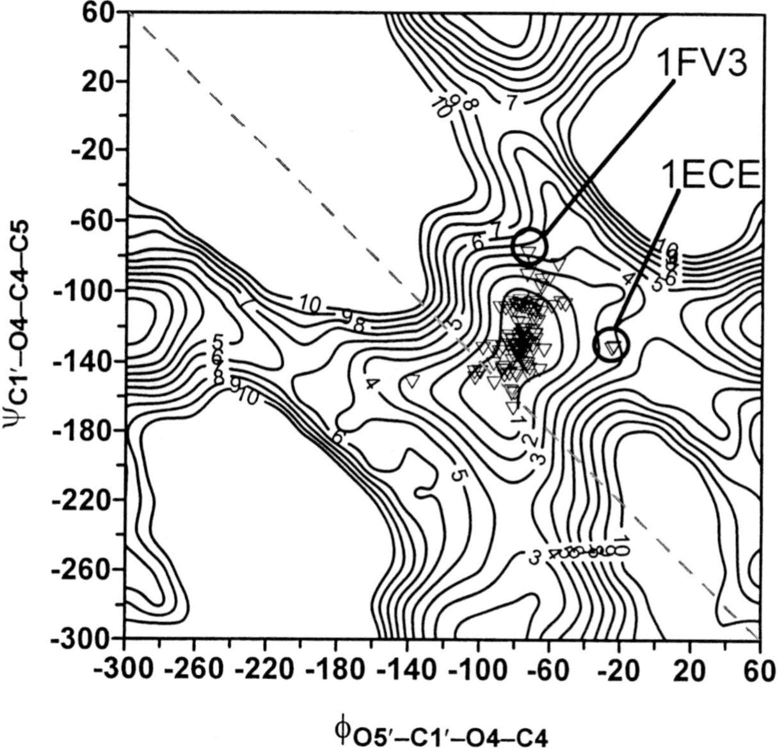

Figure 5. The same contours as in Figure 4 but with the conformations from carbohydrates complexed with proteins shown as triangles. The circled triangles are for the conformations of a disaccharide linkage in the tetanus toxin protein complex 1FV3 and the endocellulase enzyme complex, 1ECE.

Figure 5 shows the same energy surface as in Figure 4, but instead with triangles that indicate the conformations reported in crystalline complexes of proteins and carbohydrates resembling cellulose fragments. There are considerably more structures with a greater range of geometries than for the small molecule crystals despite the majority of structures being within the 1-kcal/mol contour. One high energy value, for the 1FV3 structure (*28*), is more than 4 kcal/mol above the map's global minimum. Two nearly identical linkages for 1ECE (*12*) are also circled.

Figure 6 shows the conformations from lactose moieties taken from the Protein Data Bank. Again, the majority of the conformations are inside the 1-kcal/mol contour, but the range of conformations is considerably larger with five points outside the 5-kcal/mol contour.

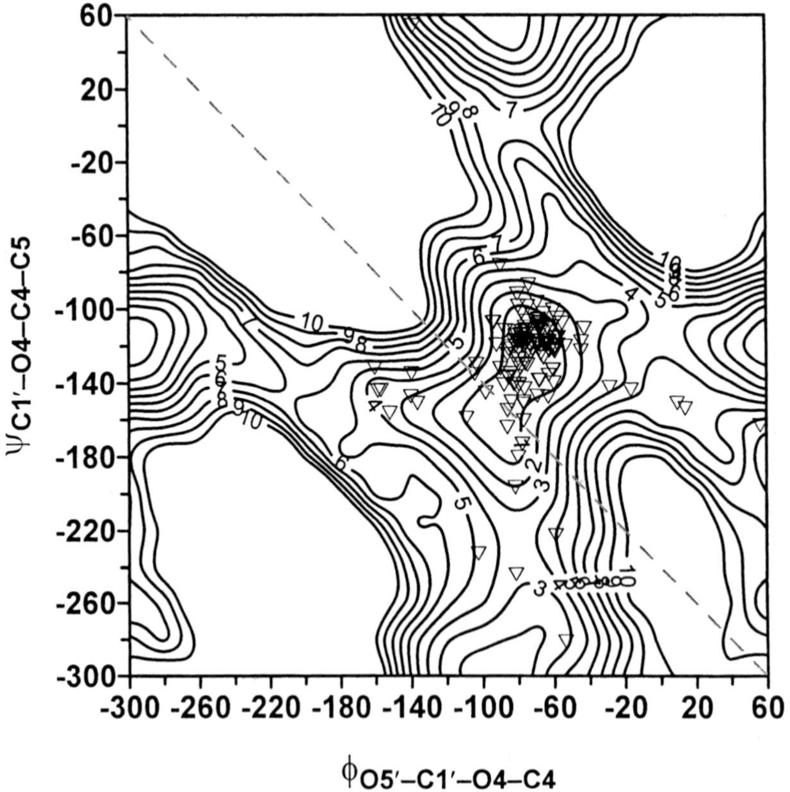

Figure 6. The same energy surface for cellobiose as in Figures 4 and 5, but this time with the crystalline conformations from complexes of lactose-related proteins.

Figure 7, upper model, shows the conformation of the 1FV3 linkage of cellobiose in a non-enzymatic protein site. The lower model shows the results of a rigid rotation to the conformation found in crystalline β-cellobiose (*11*). After the rotation, there is an implausible short distance between O3 and O5'. This suggests that there are errors in the coordinates of the residue and that the φ and ψ values are not accurately determined. Besides the variability in the energy maps based on different methods, accuracy in the crystal coordinates is the biggest obstacle in this type of work.

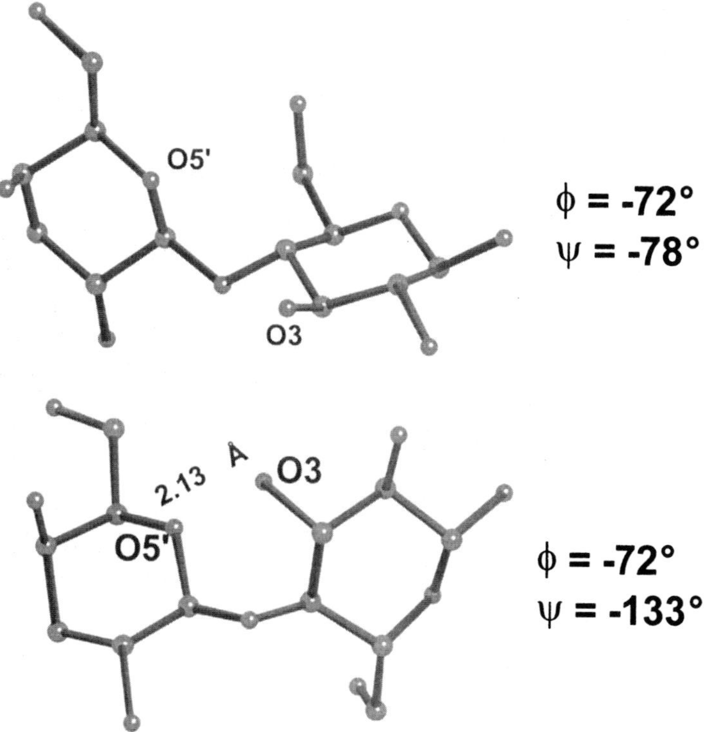

Figure 7. Cellobiose-like 3'-deoxy-β-lactose in the reported conformation from the 1FV3 protein complex (upper drawing). See also Figure 5 for the location in φ,ψ space for this linkage. The lack of the 3' hydroxyl group is not expected to substantially influence the linkage conformation. The lower drawing shows the same molecule after rotating the ψ torsion angle from -78° to -133°. There, the distance between O3 and O5' is only 2.13 Å. Since the same distance in the accurately determined cellobiose crystal is 2.77Å, the actual details of the 1FV3 are probably not well determined.
(See page 11 of color inserts.)

Conclusion

In this work we have found that scissile linkages in carbohydrate molecules are often distorted beyond the limits of distortion seen in small molecules. In some ways, this work is an update to our earlier work which had similar findings and included maltose linkages as well. Since then, we have constructed a new energy surface that is, at least in principle, more accurate, and used carbohydrate conformations from fresh searches of the Protein Data Bank.

It would be helpful if crystallographers routinely consulted applicable Ramachandran plots when analyzing their results in regard to the linkage conformations. Such consultation may assist with identification of inaccurately determined geometries, as well as characterization of the catalytic sites in future determinations.

Acknowledgements

We thank Professor Peter Reilly at Iowa State University and Dr. Oliver Dailey from the Southern Regional Research Center for comments on a draft of the manuscript.

References

1. Fischer, E. *Ber. Dtsch. Chem. Ges.* **1894**, *27*, 2984–2993.
2. Koshland, D. E., Jr. *Proc. Natl. Acad. Sci. USA* **1958**, *44*, 98–104.
3. Bosshard, H. R. *News Physiol. Sci.*, **2001**, *16*, 171–173.
4. Smith, J.; Cusack, S.; Poole, P.; Finney, J. *J. Biomol. Struct. Dyn.* **1987**, *4*, 583–538.
5. Alexandrov, V.; Lehnert, U.; Echols, N.; Milburn, D.; Engelman, D.; Gerstein, M. *Protein Sci.* **2005**, *14*, 633–643.
6. Phillips, D. C. *Sci. Amer.* **1966**, *215*, 78–90.
7. Dafforn, A; Koshland, Jr., D. E. *Proc. Nat. Acad. Sci. USA*, **1971**, *68*, 2463–2467.
8. Sasisekharaan, V. *Collagen Proc. Symp. Madras, India*, **1962**, *1060*, 39–78.
9. Rao, V.S.R.; Sundararajan, P. R.; Ramakrishnan, C.; Ramachandran, G. N. In *Conformation in Biopolymers*. Academic Press: London, **1963**, *Vol. 2*.
10. Allinger, N. L.; Chen, K.; Lii, J.-H. *J. Comput. Chem.* **1996**, *17*, 642–648.
11. Chu, S. S. C.; Jeffrey, G. A. *Acta Crystallogr. Sect. B* **1968**, *24*, 830–838.
12. Sakon, J.; Thomas, S. R.; Himmel, M. E.; Karplus, P. A. *Biochemistry* **1996**, *35*, 10648–10660.

13. Berman, H. M.; Westbrook, J.; Feng, Z.; Gilliland, G.; Bhat, T. N.; Weissig, H.; Shindyalov, I. N.; Bourne P. E. *Nucleic Acids Research*, **2000**, *28*, 235–242.
14. French, A. D.; Johnson, G. P. *Cellulose* **2004**, *11*, 5-22.
15. Hess, K.; Trogus, C. Z. *Physikal. Chem. Bodenstein-Festband* **1931**, *11*, 385-391.
16. Zugenmaier, P. In *Cellulose: Structure, Modification and Hydrolysis*; Young, R. A.; Rowell, R. M.; Eds.; John Wiley and Sons: New York, **1986**, pp 221-245.
17. Sakon, J.; Irwin, D.; Wilson, D. B.; Karplus, P. A. *Nat. Struct. Biol.* **1997**, *4*, 810-818.
18. Miyake, J.; Kurisu, G.; Kusunoki, M.; Nishimura, S.; Kitamura, S.; Nitta, Y. *Biochemistry* **2003**, *42*, 5574-5581.
19. Dowd, M. K.; French, A. D.; Reilly, P. J. *Carbohydr. Res.* **1994**, *264*, 1-19.
20. Tvaroška, I.; Bleha, T. *Adv. Carbohydr. Chem. Biochem.* **1989**, *47*, 45-123.
21. Barrows, S. E.; Dulles, F. J.; Cramer, C. J.; French, A. D.; Truhlar, D. G. *Carbohydr. Res.* **1995**, *276*, 219-251.
22. Ferretti, V.; Bertolasi, V.; Gilli, G.; Accorsi, C. A. *Acta Crystallogr., Sect. C*, **1984**, *40*, 531-535.
23. French, A. D.; Kelterer, A. –M.; Cramer, C. J.; Johnson, G. P.; Dowd, M. K. *Carbohydr. Res.* **2000**, *326*, 305-322.
24. French, A. D.; Johnson, G. P. *Can. J. Chem.* **2006**, *84*, 603-612.
25. French, A. D.; Johnson, G. P.; Kelterer, A. –M.; Dowd, M. K.; Cramer, C. J. *Int. J. Quantum Chem.* **2001**, *84*, 416-425.
26. Nishiyama, Y.; Langan, P.; Chanzy, H. *J. Am. Chem. Soc.* **2002**, *124*, 9074-9082.
27. Peralta-Inga, Z.; Johnson, G. P.; Dowd, M. K.; Rendleman, J. A.; Stevens, E. D.; French, A. D. *Carbohydr. Res.* **2002**, *337*, 851-861.
28. Fotinou, C.; Emsley, P.; Black, I.; Ando, H.; Ishida, H.; Kiso, M.; Sinha, K. A.; Fairweather, N. F.; Isaacs, N. W. *J. Biol. Chem.* **2001**, *276*, 32274-32281.

Indexes

Author Index

Subject Index

B

C

E

F

G

I

M

N

O

P

Q

R

S

T

U

V

W

X

Y